EFCE Event No. 290

4th International Symposium on

Loss Prevention and Safety Promotion in the Process Industries

Volume Three — Chemical Process Hazards

This volume forms the proceedings of the 8th Symposium on Chemical Process Hazards. Organised by the Institution of Chemical Engineers, held at Harrogate, England 12-16 September 1983

THE INSTITUTION OF CHEMICAL ENGINEERS SYMPOSIUM SERIES NO. 82

ISBN 0 85295 161 2

7120-3655

CHEMISTRY

PUBLISHED BY THE INSTITUTION OF CHEMICAL ENGINEERS

First edition 1983 — ISBN 0 85295 161 2

MEMBERS OF THE INSTITUTION OF CHEMICAL ENGINEERS (Worldwide) SHOULD ORDER DIRECT FROM THE INSTITUTION

Geo. E. Davis Building, 165-171 Railway Terrace, Rugby, Warks CV21 3HQ

Australian orders to:
R.M. Wood, School of Chemical Engineering and Industrial Chemistry, University of New South Wales, PO Box 1, Kensington, NSW, Australia 2033.

Distributed throughout the world (excluding Australia) by Pergamon Press Ltd., except to I.Chem.E. members.

U.K.	Pergamon Press Ltd., Headington Hill Hall, Oxford OX3 0BW, England
U.S.A.	Pergamon Press Inc., Maxwell House, Fairview Park, Elmsford, New York 10523, U.S.A.
CANADA	Pergamon Press Canada Ltd., Suite 104, 150 Consumers Rd., Willowdale, Ontario M2J 1P9, Canada
FRANCE	Pergamon Press SARL, 24 rue des Ecoles, 75240 Paris, Cedex 05, France
FEDERAL REPUBLIC OF GERMANY	Pergamon Press GmbH, 6242 Kronberg-Taunus Hammerweg 6, Federal Republic of Germany

Library of Congress Cataloging in Publication Data
Main entry under title:

Loss prevention and safety promotion in the process industries, 4.

(The Institution of Chemical Engineers symposium series no. 80-82)
"EFCE event no. 290"
Contents: v.1. Safety in operations and processes — v.2. Hazardous chemicals and liquefied gases — safe transport by sea — v.3. Chemical process hazards.
1. Chemical engineering — Safety measures — Congresses.
I. Institution of Chemical Engineers (Great Britain)
II. Royal Institution of Naval Architects.
III. Series: Symposium series (Institution of Chemical Engineers (Great Britain)); no. 80-82
TP149.L684 1983   660.2'804   83-17452
ISBN 0-08-031396-5 (set)
ISBN 0-08-030291-2 (Volume 1)
ISBN 0-08-030292-0 (Volume 2)
ISBN 0-08-030293-9 (Volume 3)

COMMITTEES

Symposium Chairman

Dr. W.L. Wilkinson

Executive Committee

Prof. C. Hanson *(Chairman)*

Mr. J. Borland	Mr. M. Kneale
Mr. R.C. Gray	Dr. D.A. Lihou
Mr. B.R. Harris	Mr. V.C. Marshall
Dr. P.J. Heggs	Mr. A.J. Moyes
Mr. T.A. Kantyka	Dr. P.E. Preece

Scientific Programme Committee

*Dr. H.I. Joschek *(Chairman)*
*Prof. A. Lefebvre *(Belgium)*
*Dr. Ing. J. Skarka *(Czechoslovakia)*
*Mr. N. Hjorth *(Denmark)*
*Mr. P. Hypponen *(Finland)*
*Mr. T. Lainto *(Finland)*
*Mr. R. Grollier-Baron *(France)*
*Mr. Y. Verot *(France)*
*Dr. F.W. Schierwater *(Germany)*
*Prof. Dr. H.G. Schecker *(Germany)*
*Prof. Dr. A. Barbieri *(Italy)*
*Prof. Ing. G. Ferraiolo *(Italy)*
*Mr. P.L. Klaassen *(Netherlands)*
*Dr. Ir. H.J. Pasman *(Netherlands)*
*Mr. E.N. Bjordal *(Norway)*
*Dr. H. Frostling *(Sweden)*
*Mr. A. Jacobsson *(Sweden)*
*Prof. Dr. G. Gut *(Switzerland)*
*Dr. P. Grimm *(Switzerland)*
*Mr. T.A. Kantyka *(UK)*
*Dr. J. Bond *(UK)*
*Mr. R.C. Mill *(USA)*
*Mr. Parker Peterson *(USA)*
 Dr. J.H. Burgoyne *(N.W. Branch IChemE)*
 Mr. J.A. Hancock *(N.W. Branch IChemE)*
 Mr. M. Kneale *(N.W. Branch IChemE)*
 Mr. T.R. Farrell *(The Royal Institution of Naval Architects)*
 Mr. R.C. Gray *(The Royal Institution of Naval Architects)*

*Member of the EFCE Working Party on Loss Prevention

CONTENTS

Page

PREFACE

The fourth International Symposium on Loss Prevention and Safety Promotion in the Process Industies was held in Harrogate, England, in September 1983, organised by the Institution of Chemical Engineers (IChemE) on behalf of the European Federation of Chemical Engineering (EFCE). The IChemE has always been in the forefront of endeavours aimed at the promotion of improved standards of safety in the design and operation of plants in all sectors of the process and allied industries. In fact, it was a national symposium on Loss Prevention organised by the Institution in Newcastle-upon-Tyne in 1971 which led to greater international collaboration in this field by the establishment of the EFCE Working Party on Loss Prevention. It was therefore particularly appropriate that the United Kingdom should have been selected as venue for the latest symposium to have been co-ordinated by this Working Party and the IChemE welcomed the opportunity to act as sponsor. For the first time in such meetings, considerable attention was devoted to the safe transport of hazardous chemicals and liquefied gases by sea, and the best design of the necessary shore installations, and these sessions were developed jointly with the Royal Institution of Naval Architects (RINA).

With some ninety papers contributed by experts from many parts of the world, the Symposium provided a vast amount of information, applicable to both design and operation, based on research and practical experience. To make this knowledge available to the widest possible audience and to give a permanent record, the papers are being published as the Proceedings of the Symposium. These also include summaries of the main points raised during the discussion periods which were a feature of every session. Because of the number of papers involved, it has been necessary to publish the Proceedings in three volumes, reflecting the three main sub-divisions of the Symposium itself. The first volume includes papers on the assessment of risks and effects, hazard identification, release rates of liquefied gases, and consideration of the magnitudes of fires and explosions. These are supported by a number of case histories. The second volume is devoted to those sessions organised in collaboration with the RINA on the safe transport of hazardous chemicals and liquefied gases by sea. The third, Chemical Process Hazards, incorporates the latest in a series of meetings on that theme organised by the North Western Branch of the IChemE. Topics include runaway reactions, unstable products, explosion hazards and relief.

In presenting these Proceedings of the Symposium, I would like to express sincere thanks to all those whose efforts made the event possible - to the members of the Scientific Committee for the selection of papers and development of the technical programme, members of the Executive Committee for advice and practical help with the logistics of such a major meeting, the Yorkshire Branch of the IChemE for local support, the staff of the IChemE for all their efforts, to the authors for such stimulating papers, session chairmen, and many others. All have played a vital role. It is impossible to mention individually all who have helped. However, reference must be made to

the efforts of Miss Fiona Dendy, Mr. Bernard Hancock and Mrs. Gillian
Nelson all of the Institution's staff, and to Dr. P.J. Heggs of the
Yorkshire Branch. They made exceptional contributions and gave much
support over the whole period of organisation. Finally, I would like
to pay tribute to the work of the Vice-Chairmen of the sessions who
have been responsible for summarising the various discussions.
Without their efforts we would have been denied any record of that
valuable part of the Symposium.

It is my sincere hope that the exchange of information and
experience made possible by the Symposium will play some part in the
promotion of improved standards of safety and loss prevention in our
industries. If this hope is realised, then the efforts of all those
who have been involved will have been justified.

Carl Hanson
Chairman, Executive Committee
October 1983

Safety in Operations and Processes

Chairman

W.L. Wilkinson,
BNFL, UK

THERMOCHEMISTRY OF EXOTHERMIC DECOMPOSITION REACTIONS

TH. GREWER and E. DUCH

HOECHST AG, 6230 FRANKFURT/M 80, GERMANY

The energy released by an exothermic decomposition is one of the most important quantities related to the safe handling of the substances concerned. From experimental results typical decomposition energies per mole can be attributed to some frequently occurring molecular groups, e. g. hydrazo, nitro or nitroso groups. For simple cases these quantities can be compared with thermodynamically calculated heats of decomposition. There are, however, many organic compounds for which the exothermic decomposition cannot be localized in the molecule.

The self heating of a substance may result in a real thermal explosion only if the energy of the exothermic decomposition exceeds a certain order of magnitude (about 150 J/g).

INTRODUCTION

With rising temperature every substance will at some point reach its limit of stability. There is a temperature range in which pyrolysis reactions set in that can lead to smaller molecular units, but also to larger molecules. In many cases these pyrolysis reactions, which we will collectively term "decomposition", are exothermic. If a decomposition reaction is highly exothermic, it constitutes a potential danger, but it is not easy to say what "highly exothermic" means. This is one of the questions that will be dealt with in this paper.

The amount of energy released by decomposition reactions ranges from very low values that are irrelevant to safety questions, through intermediate values to very high values that are characteristic of explosive substances. In many cases the chemical structure of a substance, usually organic compounds are involved, will reveal the class to which the substance belongs. The rules that apply here, which have been learned from both thermodynamics and experience, are a further topic of this paper.

1. Thermodynamic calculation of the energy of decomposition

In principle it is true that the energy of decomposition can be calculated from tabulated thermodynamic data, but in reality it is usually not possible because neither the enthalpies of formation of the substances nor the reaction products are known.

Table 1: Examples of decomposition enthalpies calculated on the basis of plausible assumptions on the reaction products. Measured values of the decomposition energies are given for comparison.

Substance	formula	state	heat of formation ΔH_f° kJ/mol	assumed products*	Heats of decomposition $-\Delta H_R$ calc. kJ/mol	$-\Delta U_R$ exp.(DSC) kJ/mol
Ethylene oxide	$CH_2 - CH_2$ (O)	lq	− 77,6	$CO + CH_4$	107	77
Propylene oxide	$CH_3 - CH - CH_2$ (O)	lq	− 120,7	$CO + C_2H_6$	74,5	65
1,4-Dioxane	(dioxane ring)	lq	− 354	$2CO + 2CH_4$	17	17
Phenylhydrazine	$-NH-NH_2$	lq	+ 141,8	benzene $+ \frac{2}{3} NH_3 + \frac{2}{3} N_2$	90	71
1,2-Diphenylhydrazine	$-NH-NH-$	c	+ 220,6	biphenyl $+ \frac{2}{3} NH_3 + \frac{2}{3} N_2$	69	81
Azobenzene	$-N = N-$	c	+ 320,5	biphenyl $+ N_2$	138	146
1,3-Diphenyltriazene	$-N=N-NH-$	c	+ 331,6	$-NH-$ (carbazole) $+ N_2$	214	266

* gaseous state unless otherwise stated

One of the rare examples where both the formation enthalpy
and the course of the decomposition reaction are known is
that of hydrogen peroxide, which clearly decomposes in
accordance with the reaction equation

$$H_2O_2 \longrightarrow H_2O + {}^1/_2\, O_2.$$

The enthalpy of decomposition is in this case equal to the
difference between the enthalpy of formation of water and
that of hydrogen peroxide [1] - 286 kJ/mol + 188 kJ/mol =
- 98 kJ/mol.

There are other examples of substances for which the enthalpy
of formation but not the reaction products are known. Frequently it is possible to make plausible assumptions concerning the reaction products. Some decomposition reactions of
this type are summarized in table 1.

For the vast majority of substances that decompose exother-
mally, however, neither the formation enthalpies nor the reaction products are known. We want to devote special attention
to these substances.

2. Measurement of the energy of decomposition

Most of the methods now used to measure the energy of decomposition are thermoanalytic methods, variants of DTA or DSC. In
addition adiabatic methods are used, such as ARC. The
classical method employing the bomb calorimeter is used
chiefly for substances with high energies of decomposition
(explosive substances). Wherever our own measurements of the
energies of decomposition are referred to in this paper, they
were made by a thermoanalytic method (TA 2000 by Mettler).
When we refer to emperical values we speak of "energies" of
decomposition and not "enthalpies", because they have all been
measured at constant volume and not at constant pressure.

The reaction energies measured by thermoanalytic methods such
as DSC must be considered with some scepticism, because in the
first approximation they have been determined isothermally,
whereas under practical conditions decomposition corresponds
more closely with the adiabatic limiting case. This explains
why for components with higher energies of decomposition, such
as nitro compounds, the energies of decomposition measured by
DTA or DSC turn out to be too small. This fact has been pointed
out by Yoshida [2].

That such different values are found is understandable when one
considers that decomposition often occurs in several steps,
which are visible on the DTA diagram as a succession of exothermic peaks. Figure 1 shows such a DTA diagram for a nitro
compound. The two or three peaks in the diagram do not total
up to the 400 kJ/mol that is typical of nitro compounds. But
since the DTA measurement was discontinued at 450 °C, it is
possible that at higher temperatures further peaks would
appear, because the temperature rise occurring under adiabatic

Table 2: Ranges of experimentally determined decomposition energies of organic compounds containing characteristic molecular groups

Type of compound	Group	Range of measured decomposition energies $-\Delta U$ kJ/mol	Number of substances evaluated	References
Nitro	$\diagup C-NO_2$	220 ... 410	30	4, 5
Nitroso	$\diagup C-N=O$	90 ... 290	4	5
Oxime	$\diagup C=N-OH$	170 ... 230	2	5
Isocyanate	$-N=C=O$	20 ... 30	3	5
Azo	$-N=N-$	100 ... 180	5	5, 6
Hydrazo	$-NH-NH-$	65 ... 80	3	5
Diazo	$\diagup C-N\equiv N \oplus$	130 ... 165	5	5, 7
Peroxi	$\diagup C-O-O-$	200 ... 340	20	8
Epoxi	epoxide	45 ... 80	3	5
Double bond	$\diagup C=C \diagdown$	40 ... 90	6	5

conditions amounts to more than 1000 K.

Table 1 shows the calculated energies of decomposition for a
few substances in comparison with the empirical values (DSC).
The calculations were made under plausible assumptions concerning the reaction products, and the heat of formation is
known for these substances. In some cases the agreement is not
bad, which can partly be attributed to the fact the reaction
equation was chosen to fit the circumstances.

3. Relationship between chemical structure and energy of decomposition

3.1. Assignment of energies of decomposition to groups of atoms

For certain frequently occurring types of compounds it is
possible to assign a characteristic energy of decomposition to
the group of atoms responsible for exothermic decomposition.
Table 2 indicates for a few groups of atoms the range in which
the energies of decomposition lie in several compounds of the
same type. The values listed refer to one mole of the respective atomic group, i. e. for a dinitro compound, for example,
the energy of decomposition given is half the amount referred
to a mole of the substance.

The fact that some of the energies of decomposition are spread
over a relatively wide range is due to the causes previously
mentioned: for the nitro compounds the "heats of explosion"
determined in the bomb calorimeter on substances with two or
more nitro groups lie in the neighbourhood of 400 kJ/mol [4] ,
while the lower values (between 200 and 300 kJ/mol) were
derived from DSC measurements on mononitro compounds. The
values measured with the bomb calorimeter lie in the very
narrow range of 380 to 410 kJ/(mol NO_2 group) and appear to
us to be more credible than the values measured with DTA or
DSC. If the molecular weight is not too high, the heat of
reaction is sufficient to raise the temperature by 1000 or
more K, which extends into a temperature range where thermodynamic equilibrium can already be achieved. This requires,
however, that a closed container be used. Mononitro compounds
in an open container will probably release only the smaller
amounts of energy of decomposition around 200 to 300 kJ/mol.

The data on peroxides were obtained by our colleagues Meister
and Winter, Höllriegelskreuth. The range reported includes
alkyl peroxides, diacyl peroxides, percarbonates and peresters.
However, there are no really significant differences between
these types of compounds in respect of energies of decomposition, that is to say they are all spread over roughly the same
range. The differences are due more to the different methods -
here again DSC and calorimetry - than to the differences
between types of compound.

In substances with a double bond between carbon atoms exothermal heat of reaction can arise in two ways. The first is
through true decomposition, that is the splitting of molecules,
in which pyrolysis products such as elemental carbon are formed.
The other possibility is polymerization. For the polymerization
of olefins and other compounds with double bonds there are
tabular values of the polymerisation enthalpies. For polymerization which starts in the liquid state the values given are
about 70 to 90 kJ/mol [9] . Let us formulate as an example of
decomposition a reaction for α-olefins, which probably yields
the highest possible exothermal heat of reaction:

$$C_nH_{2n+1}-CH=CH_2 \ (lq) \longrightarrow C_{n+1}H_{2n+4} \ (g) \ + \ C$$

The reaction ethalpy $-\Delta_RH$ calculated from the tabulated
thermodynamic data for hydrocarbons [1, 3] is 70 to 80 kJ/mol.
Exothermic heat of reaction can also be generated in reactions
in which aromatic compounds are formed with, for example, H_2
being split off, because of the resonance energy of these
compounds, for instance in the reaction

which is exothermic by - 24 kJ/mol if the enthalpies of
formation [3] are correct.

For some of the types of compounds listed in table 2 the
number of substances whose energy of decomposition has been
measured is small and the values given are correspondingly
uncertain. This should be kept in mind if these data are used
in safety calculations. At any rate it is advisable to use the
upper limit of the range of scatter shown in the table.

3.2. Decomposition reactions based on unspecifiable groups of
     atoms

Table 2 refers to types of compounds whose exothermic decomposition has long been known. Here the exothermic reaction is
localized in certain groups of atoms that can be located in
the molecule. Experiments have shown,that there are many other
classes of substance which exhibit the property of exothermic
decomposition, but for which the group of atoms responsible
cannot easily be specified.

One such class of substances is the organic compounds that
contain only the elements C, H and O. Table 3 summarizes the
energies of decomposition of some C-H-O compounds and from
this table it is apparent that such compounds can exhibit quite
considerable exothermic effects. Actually, however, these
reactions have long been known. The carbonization of wood is
a well-known exothermic process [10]. On the other hand there
are no signs that the exothermic decomposition of carbohydrates
can be dangerous. Hence we will return to table 3 later when
we discuss the question of evaluating the risks caused by

exothermic reactions. Because of the widely varied molar masses of the substances in table 3, the energies of decomposition given are referred to the mass unit, which is more important for the evaluation of safe practice than the values referred to the mole.

For the C-H-O compound there is no apparent systematic relationship between structure and energy of decomposition. In these compounds the energy-generating process may be considered to be the conversion of O-C bonds to O-H bonds. Because of the higher binding energy of the O-H bond, an amount of energy equalling at most 40 to 50 kJ /mol can be released here.

We now come to the substances in which it is not a single group of atoms but two jointly acting groups of atoms that are thought to be responsible for decomposition. We have previosly referred to sucn combinations of atomic groups, which can jointly cause exotnermic decomposition [11] . We will consider here the example of the aromatic chloro-amino compounds. For the chloro-amino derivatives of benzene the temperature at which aecomposition begins depends on the relative positions and the number of chlorine atoms and amino groups. How the energy of decomposition depends on the position in the molecule is shown for the simplest case of the chloroaniline isomers in table 4. The highest energy of decomposition is found witn tne p-compound, it is smaller in the o-compound and the m-compound exhibits no exothermic effect at all.

4. Evaluation of the hazard of exothermic decomposition

We now come to the most important but at the same time most difficult chapter of this paper. It concerns the question: can one judge from the energy of decomposition whether a substance is dangerous or not? If the answer is "yes" there must be a limiting value for the energy of decomposition below which the exothermic reaction is no longer relevant to safety engineering. We cannot yet answer this question fully, but we can perhaps provide information which is of same practical benefit.

To evaluate the violence of a decomposition reaction it is not the energy released per mole but the energy per mass unit that is decisive. The energy of decomposition in J/g is closely related to the adiabatic temperature increase ΔT_{ad}. Since the specific heat of organic substance usually lies in the relatively narrow range from 1.5 to 2 J/g.K, we can assume that ΔT_{ad} is roughly proportional to the energy of decomposition in J/g.

In table 5 we have compiled the energies of decomposition in J/g for a few arbitrarily chosen substances. They are arranged in the order of decreasing energy of decomposition. At the top are explosive substances with very high energies of decomposition of a few thousand J/g. Substances which cannot detonate but are nevertheless dangerous in chemical plants lie in the range from 500 to 3000 J/g. Below 500 J/g we find such harmless

Table 3: Decomposition energies of C-H-O compounds
(Apparatus: Mettler TA 2000, sample container Cr-Ni,
3 K/min)

Substance	Temperature region of decomposition °C	Decomposition energy $-\Delta U$ kJ/mol	J/g
Ethylene oxide	300 ... 480	77	1740
1,4-Dioxane	130 ... 200	17	165
Glucose	220 ... 330	73	406
Sucrose	230 ... 350	165	480
Cornstarch	260 ... 340		460
Cellulose	200 ... 300		330
Polyvinylalcohol	125 ... 430		540

Table 4: Decomposition energies of the isomer chloroanilines
(Apparatus: Mettler TA 2000, sample container glass,
3 K/min)

Substance	Temperature region of decomposition °C	Decomposition energy $-\Delta U$ kJ/mol
2-chloroaniline	210 ... 400	52
3-chloroaniline	-	0
4-chloroaniline	200 ... 340	81

Table 5: Decomposition energies $-\Delta U$ of selected substances,
in the order of decreasing values in J/g.

Trinitrotoluene	5 100
Dinitrobenzene	4 600
Nitrobenzene	$\sim$ 3 000
Phenyldiazonium chloride	1 500
Dibenzoylperoxide	1 390
4-Nitrosophenol	1 200
Azobenzene	800
Phenylhydrazine	660
4-Chloroaniline	630
Sucrose	480
1,2-Diphenylhydrazine	440
2-Chloroaniline	410
Cellulose	330
1,4-Dioxane	165

substances as saccharose and cellulose. Thus the limit that
suggests itself for the hazardousness of exothermic decomposition is in the region of 500 J/g. However, we have to add
the qualification that the substance must not be in a completely "closed" container.

Consequently we must apply different standards depending on
whether the substance is in an "open" or a "closed" container.
The terms "open" and "closed" can refer only to a quantitative
difference, since a truly closed container can be made to
burst without an exothermic reaction. If we classify an "open"
container as one with an opening the size of a manhole, for
example, and a "closed" container as one with merely a safety
valve, we can propose the following limiting values:

"open" (opening/volume $> 10^{-1} m^{-1}$) about 500 J/g
"closed" (opening/volume $> 10^{-3} m^{-1}$) about 150 J/g

The characteristic opening/volume ratios are somewhat arbitrary,
but we think that the order of magnitude is correct. A more
exact analysis which we do not attempt here would show that not
the reaction energy itself or the adiabatic temperature rise
ΔT_{ad} is the relevant quantity, but the dimensionless quantity
$\mathcal{E} = E \, \Delta T_{ad}/RT_o^2$ is decisive for the development of a thermal
explosion. This means that beside the decomposition energy
also the energy of activation E will influence the hazard of
the decomposition reaction. A relatively small activation
energy will reduce the risk, and maybe this is the cause for
the low risk involved in the decomposition of the carbohydrates
in spite of their not negligible decomposition energies.

In any case, an exothermic decomposition energy below 150 J/g
for an organic substance has to be regarded as irrelevant under
safety aspects. In these cases ΔT_{ad} is so small that only a
poor acceleration of the reaction will be possible.

In many cases additional factors must be considered, such as
the existence of phase transformations or endothermic reactions.
An example is shown in figure 2 in the form of a DTA diagram.
Here an exothermic reaction follows an endothermic effect.
Since both effects are of about the same magnitude - only the
signs are different - they cancel one another so that this instance of exothermic decomposition must be considered as non-
hazardous.

This paper has been partly supported by the German Ministry of
Research and Technology. The authors thank H. Carolus for his
assistance in the experimental work.

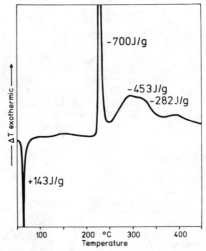

Fig.1. Quantitative DTA Diagram of
3-Nitrobenzaldehyde: Distribution
of decomposition energies on different peaks.

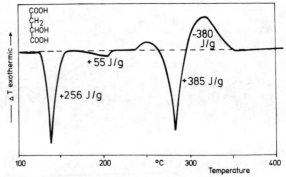

Fig.2. Quantitative DTA Diagram of Malic Acid:
Compensation of endothermic and exothermic
effects.

REFERENCES

1. B.J. Zwolinski and R.C. Wilhoit,
   Heats of Formation and Heats of Combustion, Thermodynamics
   Research Center, Texas A & M University, 1968

2. M. Itoh, T. Yoshida, M. Nakamura and K. Uetake,
   J. Industr. Explosives Soc., Japan, 38 (1977) No. 1,17

3. J.B. Pedley and J. Rylance,
   Sussex-NPL Computer Analyzed Thermochemical Data,
   Univ. of Sussex 1977

4. R. Meyer,
   Explosivstoffe, 3. Aufl., Verlag Chemie, Weinheim 1977

5. E. Duch and Th. Grewer,
   unpublished results

6. S. Morisaki, M. Naits and T. Yoshida,
   J. Hazard. Mat. 5 (1981), 49

7. P.D. Storey,
   Symposium on Runaway Reactions, Unstable Products and
   Combustible Products.
   The Inst. of Chem. Eng. Symposium Series No. 68

8. Private communication of W. Meister and H. Winter,
   Höllriegelskreuth (München)

9. R.M. Joshi and B.J. Zwolinski,
   Heats of Combustion Studies on Polymers,
   Thermodynamics Research Center, Texas A & M University, 1967

10. H.J. Heinrich and B. Kaesche-Krischer,
    Brennstoff-Chemie 43 (1962), 142

11. Th. Grewer,
    Chem. Ing. Tech. 51 (1979) 928

THERMALLY STABLE OPERATING CONDITIONS OF CHEMICAL PROCESSES

H. Fierz, P. Finck, G. Giger, R. Gygax

CIBA-GEIGY Ltd., Basle, Switzerland

SYNOPSIS

The thermal risk of a chemical production procedure depends mainly on the way the synthesis is controlled and particularly on the degree of loss of its control in case of a cooling failure. We advocate here to make minimisation of this risk one of the goals of process development and present an experimental technique established in our company which favours a data-oriented approach to such risk-conscious process design.

The method is illustrated by examples.

INTRODUCTION

A risk specific for the industries handling and processing chemical materials stems from the tendency of these materials to undergo chemical transformations. These are intrinsically accompanied by an exchange of heat, mostly an evolution of heat towards the surroundings. If this heat cannot be dissipated it presents a potential danger, mostly by causing a temperature rise and thereby a possibly sharp pressure increase.

APPROACH TO RISK ASSESSMENT OF THERMOCHEMICAL HAZARDS

Physical operations

For physical plant operations such as drying, grinding etc. and for storage and transportation the occurrence of chemical transformations is utterly unwanted. Safety testing has therefore focused strongly on the detection of exotherms which are considered indicative of conditions to be avoided. The goal of such an approach is the definition of safe operation temperatures which are sufficiently low to exclude the triggering of any chemical reactions under some given conditions. The chemical stability of a compound or mixture is often

treated like a property which can be measured and assigned to the
materials investigated.

Chemical processes

For risk assessment of a desired chemical process this approach is
much too limited, for in this case one is primarly interested in running a particular chemical reaction fast and safely. The plant cooling
equipment is designed to cope with the substantial heat release rates
originating from the desired synthesis. Under normal conditions any
undesired secondary reactions typically proceed much slower than the
desired synthesis, and their heat evolution rates can be neglected in
comparison.

The most important deviation from normal conditions in our context
is the loss of an adequate cooling capacity. Given the comparatively
high rate of the synthesis reaction at the process temperature, it is
obviously also this reaction which determines the temperature course
in the first instant of a cooling failure. Unless the reaction can
be stopped by some means it will cause the temperature to rise, eventually reaching ranges at which consequential secondary effects become
dominant (1).

Analysis of typical plant accidents show that very often the loss of
control of the desired reaction is the triggering event. The runaway
of the moderately exothermic desired reaction by itself often is of
limited vigour due to reagent depletion. However, by rising the temperature it enhances the chance of a much more destructive potential of
a follow-up reaction to be released (Fig. 1).

Speaking in the terms of risk analysis, the desired reaction is con-
tributing predominantly to the probability of an unwanted event while
the major factor determining its severity frequently stems from follow-
up decompositions.

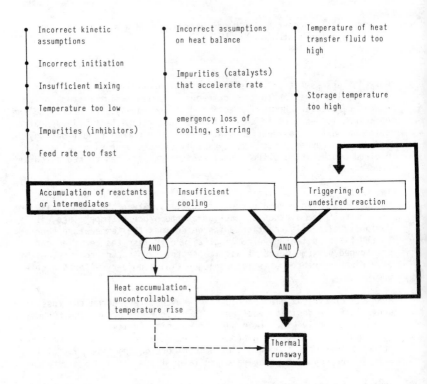

<u>Figure 1:</u> Causes of runaways in industrial reactors and storage tanks

INTEGRATED PROCESS DESIGN FOR
SAFETY OF CHEMICAL REACTIONS

As compared to the risk of physical operations the necessity of a shift in emphasis when dealing with desired reactions is implicated by the above discussion (2-5):

Instead of measuring and providing passive protection against an inherent risk of a given process, the goal is to primarily minimize this risk by actively redesigning a process with safety in mind. In addition to the thermo-kinetic properties of the chemical materials on the one hand and the plant conditions on the other, the <u>process design parameters</u> enter as a third element determining the safety of the process. Examples are the temperature profiles, the sequence and the rates of addition of individual compounds. Other than the first two factors these parameters allow for considerable freedom of choice which makes the optimisation of the degree of control over the synthetic reaction possible.

Prerequisites of this approach are:

1. Enough experimental data must be available to permit an educated
   and safety-conscious choice of the process parameters. They should
   allow to draw a rough scenario of thermal events to be expected not
   only for the regular case but also for possible emergency cases.

2. The consideration of the safety aspects of a synthesis necessarily
   has to take place at its development stage, and ultimately it
   should become an integrated part of process development.

More often than not the proposed data-oriented approach to process
design also favours other, e.g. economic goals as it makes possible
an integrated optimisation with respect to all the important goals.
Moreover it assures an adaptive scale-up without necessitating new
measurements at each stage.

EXPERIMENTAL PRINCIPLE

The instantaneous heat evolution rate of chemical and physical processes proves to be an ideal property to measure. Not only is it indicative of whether and how fast reactions are actually occuring but it
is also directly the risk-related quantity which can be easily transformed into a temperature increase rate for the case of emergency loss
of cooling. An instrument applicable for integrated process design
should, therefore, allow to reproduce and vary the parameters defining
a plant procedure and at the same time measure the instantaneous heat
release rates. The Ciba-Geigy Bench Scale Calorimeter (BSC, fig. 2,
ref. 6), which has been in use mainly by reaction development depart-
ments, fulfills these requirements. It is an active heat flow calorimeter equipped with all the modules familiar from the plant reactors
such as stirrer, feed controller, reflux condenser etc. It consists
of a standard two liter glass reactor surrounded by a jacket in which
a heat transfer fluid is circulated at a very high rate. A cascaded
controller adjusts the temperature of the circulation loop in such a
way that the heat transferred through the reactor wall and the heat
produced always equilibrate. The heat flow through the wall is a function of -1- the temperature difference actively established between
the jacket and the reactor and -2- a calibration factor containing the
heat transfer coefficient and the wetted heat transfer area. The
second factor varies with the reactor content and stirring conditions.
The calibration factor is, therefore, repeatedly determined during the
measurement by adding a known heat input rate on top of the experimental heat flux by means of an electric calibration heater. Thus
while all the external conditions correspond to the plant process
description, on-line measurement of the actual heat evolution rate is
available at all times. Firstly, this measurement provides direct information on how much heat must be expected in the plant. Secondly, by
comparing the heat evolution dynamics relative to temperatures and
feed rates it allows to draw conclusions on the degree of accumulation
of reactants as a function of process time.

This point will now be further illustrated by way of some examples.

Figure 2: The Ciba-Geigy Bench Scale Calorimeter

EXAMPLES

1. Nitration reaction

The introduction of a second nitro-group into a substituted nitro-
benzene is known to proceed sluggishly and has heretofore been performed by slowly feeding a large excess of mixed acid to a solution of
the substrate at 80 ^{o}C. Experience and microthermoanalytical data indicate a strongly exothermic, delayed decomposition to take place at
elevated temperatures.

Isothermal experiments

Fig. 3a shows a BSC trace of the reaction process starting with the
feed of mixed acid. For our purpose the feed was stopped prematurely
after an equimolar amount of mixed acid had been added. Then the unreacted species present were allowed to complete their reaction isothermally. The integral under the whole curve corresponds to -270 kJ
per kilogram of the final amount of reaction mixture. Now we will
deduce from this single experiment what can be predicted to happen
in case of an instantaneous loss of stirring and cooling at the worst
moment, i.e. when the reacting partners are present in equimolar proportion:

a) The heat flux at this moment amounts to about 12 W/kg corresponding
   to an initial temperature rise rate of ca. 25 ^{o}C per hour. Under
   adiabatic conditions a runaway within the time scale of an hour or

A16

so can be deduced from this data.

b) The overall heat evolved after equimolar addition, i.e. the integral under the heat flux curve from this moment to the completion of the reaction amounts to -180 kJ/kg. This heat arises from reagents unreacted at the considered moment and will inevitably be set free causing a temperature rise by 110 °C assuming absence of cooling. Thus in the adiabatic case, a condition which is not unrealistic for a large unstirred system, a temperature of 190 °C will be reached due to runaway of the synthesis reaction. From other sources it is known that the dangerously exothermic decomposition reaction will immediately continue the runaway at this temperature.

While a risk assessment focusing too much on the high potential of the decomposition reaction will likely tend to lower the upper limit of the "allowed" temperature range, clearly the opposite has to be done to improve the safety of the process.

Fig. 3b gives a BSC-experiment analogous to the above, only this time the reaction is run at 100 °C. The maximum temperature rise predicted in case of an instant switching from controlled to adiabatic conditions this time is on the order of 40 °C, thus a final temperature of 140 °C is anticipated. This temperature is stable for hours and enough time is available to take measures avoiding the eventual runaway of the decomposition reaction.

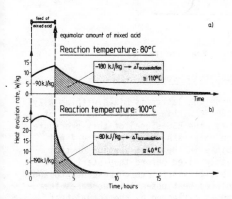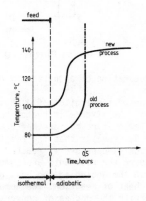

Fig. 3: Nitration
        isothermal experiments

Fig. 4: Nitration
        adiabatic experiments

Adiabatic experiments

By controlling the jacket temperature of the BSC such as to avoid any temperature gradient across the wall adiabatic conditions can be simulated. The two experiments above were rerun with the instrument being switched to the adiabatic mode at the point of equimolar addition

(Fig. 4). The above predictions are corroborated: while the 80 °C-procedure results in a steep runaway phase which had to be interrupted in the laboratory for obvious reasons, the 100 °C-procedure after an accelerating phase levels off near 140 °C.

2. Diazotisation

The optimised version of the diazotisation process of Fig. 5 shows that by selecting proper conditions a feed controlled reaction can nearly be attained. With the first drop fed the heat flux signal assumes the full value which is approximately constant during the linear feed. With the cut of the feed the signal quickly returns to the baseline. For the synthesis of diazo-compounds, this is important as some of them decompose significantly at temperatures only slightly above the diazotisation temperatures.

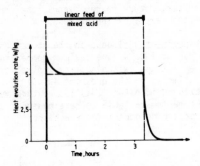

Fig. 5: Diazotisation - feed controlled

3. Preparation of a quaternary phosphonium compound

For the quaternisation of a phosphine the latter is mixed with an alkyl halide. The mixture is heated to 160 °C within two hours and held at this temperature for another 6 hours. The BSC-signal of this procedure (Fig. 6a) reveals that the reaction only takes place once the final reaction temperature is reached. Since in the batch version of the procedure the reactants are completely mixed before they start to re-act, in the beginning the heat potentially set free corresponds to the total heat of reaction. The accumulated heat potential decreases there-after with the degree of conversion as reflected by the actual heat flux curve. The maximum attainable temperature traced in the lower half of Fig. 6a is calculated using the area under the heat flux curve from the considered moment till total conversion and the heat capacity of the material.

In Fig. 6b and c, the same procedure is repeated this time feeding the phosphine within 4 and 8 hours, respectively. The actual heat re-leased at any moment can be compared to the hypothetical regular heat evolution rate (dotted line) reflecting ideal feed-controlled dynam-ics. From the difference of the areas under the ideal and real curves the maximum attainable temperature can, again, be estimated. Here it is assumed that the feed is immediately stopped in case of cooling failure, a requirement which can be implemented easily in the plant. By selecting the feed rate the adiabatically attainable temperature can practically be tuned to an acceptable value.

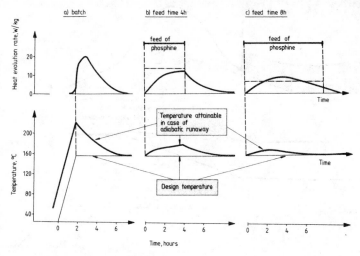

Fig. 6: Quaternisation of a phosphine - influence of addition rate

4. Sulfonation

The sulfonating agent is fed to a solution of the substrate at 60 °C. Then the temperature is raised within 5 hours to 110 °C.

Fig. 7 gives the BSC trace corresponding to this process. During the feed at 60 °C a constant heat flux is measured which corresponds to the heat of mixing of the reagents. The heat of reaction is set free during the slow temperature ramp. In the manner described above, the maximum attainable temperature can be deduced. As can be seen, the intended temperature of 110 °C is only slightly exceeded in the case of a cooling failure. This evidence was also confirmed by adiabatic experiments omitted here for lack of space. Although from the safety point of view a feeding process is generally considered to be much better, in this case the BSC provided evidence that this is not ultimately necessary and that a batch procedure can be tolerated.

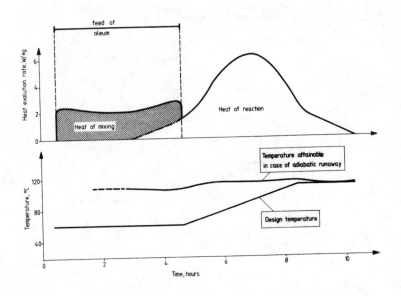

Fig. 7: Sulfonation - batch experiment

CONCLUSIONS

While in a continuous production line of a mono-plant, thermochemical properties intrinsically have to be taken into account at the design stage, the situation is much different in the speciality chemicals industry where a great variety of chemicals is produced in a more or less standard inventory of batch reactors being, essentially, scaled-up versions of laboratory glassware. Development and production has traditionally been performed by chemists rather than process engineers. This may be one of the reasons why the advantage of data-based process design procedures have not always been considered.

We wish to conclude with the hope of having indicated a simple approach to data-oriented process design taking into account, among other things, the aspects of thermal safety. This method of process development directs investments in hazard prevention to cases where they are needed most, and it avoids costly secondary measures by attacking danger at the root of its causes.

REFERENCES

1. F. Brogli & K. Eigenmann, Proceedings of the "Colloque sur la Sécurité dans l'Industrie Chimique" p. 54, Mulhouse (1978)

2. G. Giger & W. Regenass, 11th NATAS Symp. New Orleans, 1981 paper 100; and references cited therein

3. F. Brogli, G. Giger, H. Randegger and W. Regenass, I. Chem. E. Symp. Ser 68 (1981), 3/M:1

4. D. W. Smith, Chemical Engineering, Dec. 13, 1982, p. 79

5. A. Lippert & V. Pilz, Ger. Chem. Eng., 1982, 116

6. W. Regenass, ACS Symp. Ser 65 (1978), 37

Runaway Reactions

Chairman

J.H. Burgoyne
Consultant, UK

CONTROLLING THERMAL INSTABILITIES IN CHEMICAL PROCESSES

N. Schulz, V. Pilz, H. Schacke, Bayer AG, Leverkusen

INTRODUCTION AND BASIC PROBLEMS

The course and kinetics of chemical reactions are governed by physical-chemical parameters such as : temperature, pressure, concentration of reactants and residence time (or the residence time distribution). The design and equipment of the technical plant, in conjunction with adequate measurement and control devices, guarantee proper performance according to preset parameters.

If unwanted chemical reactions can occur in a plant, in particular of the kind which, by their own heat of reaction, might become self-accelerating, then special safety measures are implemented. Such measures or devices effect an additional control over process parameters (especially temperature and pressure) and, when safety limits are exceeded, counteract the cause, by, e.g. switching-off the heating systems, starting emergency cooling and adding inhibitors. Pressure relief systems might also be installed as safety measures against unwanted chemical reactions.

The following procedure is normally employed for the determination of safe operation ranges for process parameters and for the evaluation and design of safety measures and devices :

1. The possibility of an unwanted reaction (runaway reaction or thermal decomposition) has to be investigated under those conditions which are present in the normal technical environment.

Since the technical equipment, measurement and control systems and human action essentially influence and guide the course of the chemical reaction, and hence the chemical process,

2. the determination as to whether equipment failure (e.g. apparatus or control systems) could cause an unwanted chemical reaction has to be carried out. Likewise, the influence of human error on the designated process has to be taken into account.

These requirements can only be met by a successive, systematic, process-oriented procedure (screening/detailed investigation/determination of the preventive measures), in which chemists, physical-chemists and process engineers work closely together [1, 2, 3].

Since experiments cannot conceivably be performed in the scale ratio of 1:1, there exists the problem of choosing the correct experimental method; nevertheless, the results of experiments performed on the small scale must clearly be transferable to the large scale. This can best be guaranteed if the reactions are carried out under similar conditions to those present on the plant scale, as stated earlier (e.g. adiabatic or isothermal conditions present in both cases). Sometimes, however, it is only possible to perform a mathematical transformation of the experimental results, e.g. if special heat removal conditions are present.

PROBLEM-ORIENTATED EVALUATION AND SELECTION OF INVESTIGATION METHODS

The following examples will serve to illustrate the above arguments.They help to show how problem-orientated safety and loss prevention investigations are used to solve even difficult problems and to avoid accidents.

The fundamental cases given in the examples are already out of date. Modern-day conventional technical safety equipment utilized by chemical reactors would prevent most negative effects occuring, even, for example, in the case of a failure in the cooling system.

Example 1

The solvent polymerization of a resin was performed in a vented reaction vessel with a condensate reflux. The enthalpy of the reaction was not sufficient to completely vaporize the solvent and a decomposition was only measured well above the boiling point of the solvent. However, due to failure of the cooling system, decomposition occured, resulting in a pressure build-up and the explosion of the bursting disc.

What had happened ?

When the cooling system failed, the reaction mixture was at its boiling point, as always throughout the reaction, with the difference, however, that the reflux had now ceased. (The uncondensed vapour was flowing into the atmosphere). As a result, the boiling point of the monomer/polymer/solvent mixture was continually rising, because :

a) the monomer had a higher vapour pressure than the monomer/polymer mixture (vapour pressure reduction due to solids in solution), and

b) the monomer had a much higher boiling point than the solvent.

The course of the reaction is thus explicable : Owing to progressive depletion of the solvent and the decrease in the vapour pressure of the monomer/polymer mixture, the boiling point of the mixture rose to such a level that it entered the region of the undesired decomposition reaction. Since this reaction is highly exothermic and, moreover, large quantities of gas are released, the venting of the condenser was no longer sufficient for the removal of the gas and vapour flows, thus resulting in a rapid pressure build-up and the bursting of the disc.

This example demonstrates that for the estimation of hazards due to thermal instabilities, it is very important to understand as accurately as possible the progress of a disturbance by consideration of the technical data (in this case vapour pressure and thus boiling point as a function of the composition of the mixture). A simple evaluation of the form "the enthalpy of the reaction is smaller than the evaporation energy of the solvent" is not always sufficient to estimate the dangers realistically. In this example, the question concerning the method by which the decomposition reaction was detected is of secondary importance. However, the situation is different in the following example.

Example 2

The residues of a volatile compound were separated from dimers in a distillation still. It was known from sensitive DTA measurements, performed at a low heating rate, that the exothermic reaction of the dimer formation would occur at the existing temperature in the still.

This reaction was, however, tolerated, since the volatile component represented only 20 % of the whole, and because no further exothermic reaction was observed above the dimerization point. An adiabatic temperature increase of $\Delta T = 80$ K resulted from the dimerization reaction. During a malfunction of the dephlegmator in the column, the heating and vacuum were switched off. The subsequent rise in temperature in the still was accepted, since such a temperature increase always occured towards the end of a distillation. After the temperature had increased by 25 K, a steep rise in pressure suddenly ensued, accompanied by an accelerated increase in temperature. The opening of the safety valve in the column prevented the still from exploding.

What had happened ?

The cause of the temperature and pressure increases was an exothermic separation of gas from the monomers, which did not, however, occur with the dimers. This was not discovered in the DTA - measurements nor in the prior laboratory distillations, since in both of these cases the dimerization reaction had already been completed at the temperature at which the gas separation took place. For the same reason, no gas was liberated towards the end of the distillation, despite the elevated temperature.

This type of gas separation is discovered if the investigations and tests are performed in such a way as to simulate as closely as possible all of the conceivable disturbance factors to which the plant is subject. Here this would mean employing an adiabatic measuring method (if possible with pressure measurement) or at least, in addition to the first DTA-measurement, making a supplementary measurement with a high enough heating rate such that the dimerization reaction safely occurs in the temperature range produced by the adiabatic temperature rise. In this case, the gas generation reaction also occured in the laboratory investigation (see Fig. 1).

While the reactions considered in these first two examples have involved "normal" kinetics of the n-th order, the problem in the next example is concerned with more complicated kinetics.

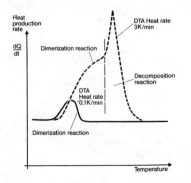

Figure 1

dimerization with subsequent decomposition (experimental results with 2 different DTA-measurements)

Example 3

A nitro-compound was usually stored for several hours at $120\,^{\circ}C$ before proceeding with the next step of the reaction. A longer storage time was necessary due to a delayed action in the following stage of the reaction. After approx. 38 hours, the temperature in the container underwent a "spontaneous" increase.

This incident was at first inexplicable, since the available DTA-measurements showed that an appreciable decomposition reaction only occured above $180\,^{\circ}C$, and even after a 24 hour adiabatic heat accumulation experiment at $120\,^{\circ}C$, a temperature rise had not been measured.

A further adiabatic heat accumulation experiment performed at $120\,^{\circ}C$, likewise resulted in the decomposition of the product after approx. 38 hours, with a temperature rise also being observed shortly before the decomposition. Further adiabatic heat accumulation tests at higher temperatures yielded shorter induction times with otherwise the same course of reaction (see Fig. 2). Evidently the decomposition was of the autocatalytic type. This decomposition mechanism can certainly be detected by means of : adiabatic heat accumulation experiments at different temperatures ; isothermal DTA-measurements ; DTA-measurements with prior isothermal storage ; or by means of an isothermal stepwise DTA-measurement (see Fig. 3).

The first indications of an autocatalytic reaction mechanism can also be obtained by performing conventional DTA-measurements, but at different heating rates and by utilizing Accelerating Rate Calorimetry ("ARC").

The above described examples naturally only represent a section of the difficulties involved in the safety and loss prevention investigations. Further problems in which the close connection between process control, physical-chemical data and technical implementation in the plant play a significant part, are :

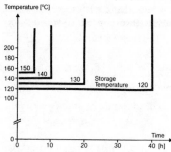

Figure 2 : temperature vs.time
for autocatalytic decomposition of
a nitric compound (experimental
results at 4 different storage
temperatures

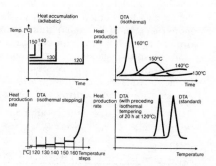

Figure 3 : detection of autocatalytic
behaviour of decomposition reaction
(various experimental methods)

- silo storage of solids, in which, owing to the extremely small heat
  removal by conduction, even infinitesimally small heat production
  rates can lead to the build-up of a decomposition reaction [4] (see
  Fig. 4). This problem can be overcome either by the extrapolation of
  measurements from adiabatic heat accumulation experiments or by
  isothermal heat accumulation in connection with a prior and a subsequent DTA-measurement or by isothermal heat accumulation in
  association with chemical analysis of the material.

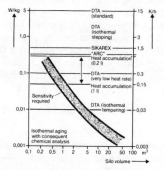

Figure 4 :

silo storage (required and
achieved sensitivity of test
methods) [4]

- reactions which are highly pressure-dependent, e.g. because a highly
  volatile component decisively influences the course of the reaction.
  In this case, comparative measurements in open and closed vessels
  would be suitable.

- heterogeneous reactions, in which re-activation of the stirrer
  mechanism can cause a potential hazard due to the accumulation of
  one of the reactants during stirrer failure. With these heterogeneous
  systems, investigations in a stirred calorimeter are often necessary,
  since the mass transfer is usually the rate determining step.

No mention has yet been made of the investigations to determine the hazards which can arise from undesired reactions with the materials of construction of the apparatus or possibly with other catalytically acting impurities. Hazards can also arise from undesired inhibitors. Investigations must, therefore, also consider :
- possible changes of reactants
- deviations in the concentration
- reactions with cooling and heating media

DTA-methods are usually appropriate and adequate to answer such questions.

INVESTIGATIONS DURING LABORATORY DEVELOPMENT

Since the investigation methods are closely linked to the conditions prevailing in the technical process, the question arises as to which investigations should be carried out when the process is still at an early stage of development, i.e. when the process parameters have not yet bee established (the so-called "base set").

The main task at this stage is to analyse the intended materials and material mixtures for the occurence of undesired exothermic reactions and for the separation of decomposition gases. Also of interest is the possibility of reactions occuring with the apparatus materials and heat transfer media.

DTA-measurements are again most frequently used for this purpose and at least 2 measurements at different heating rates should be performed for each material (e.g. 3 K/min. and 0.3 K/min.) to obtain initial indications of autocatalytic and overlying reactions. Another variant is to perform one measurement at the usual heating rate (approx. 3-5 K/min. and one isothermal stepwise DTA. The first indications of gas separatio can be found by breaking the glass DTA-investigation vessel and then by simple test-tube experiments. If gas evolution is detected, it can be investigated more accurately using Accelerating Rate Calorimetry.

EVALUATION OF PROTECTIVE DEVICES

So far only the difficulties involved in clarifying a wide variety of reactions have been discussed. The final aim of Safety Engineering is, however, to provide suitable means of preventing disturbances in plants, or at least to minimize their undesired effects. These measures help to establish safe ranges for the process parameters : temperature, pressure, time and concentration; but in some cases, special technical protective measures must be planned and evaluated.

The most important parameter in the planning of suitable countermeasures and for the design of protective devices is the heat production rate of a reaction (in W/kg) as a function of the temperature. Consequel ly, even the smallest detectable heat production rate is of significance, i.e. the detection limit or sensitivity of a process. High sensitivities are necessary, as for example in the determination of alarm and cut-off points in large vessels. The possibility of measuring the macro-kinetics of a disordered reaction course, including the heat production rate, is another factor which governs the choice of method and is useful, for example, for the evaluation and checking of the emergency cooling systems. A further selection criterion is whether there exists a possibility for the measurement of pressure and the corresponding gas flow rates. This is important for the dimensioning of ventilation pipes and pressure relief devices. Fig. 5 compares a number of investigation methods in terms of efficiency.

Effectiveness of available test methods

method	sample size	sensi- tivity	measured parameters			expense
		W/kg	pressure rise	heat pro- duction rate	kinetics	
DTA (standard)	mg	5	-		-	
DTA (isotherm stepping)	mg	0,5	-	+	(+)	
DTA (very low heat rate)	g	0,1	+	+	-	
DTA (with tempering)	mg	0,01	-	(integral)	(+)	
"ARC"	g	0,5	++	+++	++	
Heat accumulation (adiabatic)	kg	≈0,05	++	+++	+ (+)	
Reaction calorimetry	kg	0,5	++	+++	+++	

Figure 5 :

comparison of test methods

Although the described laboratory investigations will suffice in most cases, it is, however, for certain specific problems, necessary to perform pilot plant scale experiments for the safety system evaluation. Here are two examples :

Example 1

In a cyclization reaction, styrene is continuosly fed through an immersion stem into the pre-filled reaction vessel containing the other components. At the temperature prevailing in the reactor, pure styrene polymerizes while generating extreme heat. The question was whether, if the styrene was not sufficiently well mixed into the existing vessel contents, a "hot spot" could develop and possibly "ignite" the surrounding product which is thermally unstable. Laboratory scale experiments cannot provide a definite answer, since the material and heat transfer conditions existing in the full scale plant cannot be simulated on laboratory scale. Pilot plant experiments can, however, provide the necessary information. (Nevertheless, this type of experiment demands increased safety precautions.)

Example 2

Pressure release devices for reactors are necessary when an undesired exothermic reaction cannot be controlled or excluded with sufficient certainty. In cases where the vapour pressure of the reactor contents substantially exceeds the ambient pressure, effective cooling of the contents by evaporation using a pressure relief system is possible. To design such pressure release devices, not only must the reaction kinetics be known, but also it must be ascertained as to whether foaming of the reactant contents will occur with the release of pressure (evacuation of the vapour is then impeded by two-phase flow). While the kinetics can usually be measured for this purpose in the adiabatic heat accumulation test or by utilizing Accelerating Rate Calorimetry, the second problem must be solved by a pressure relief experiment performed on the pilot plant scale.

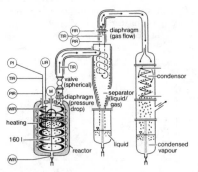

Figure 6 :

test stand for evaluation of pressure relief systems

Fig. 6 shows a test stand equipped for this purpose, as installed in our Process Engineering Department [5]. If possible, the relief tests are performed with an inhibited original substance, on to which the influence of the reaction is analytically superimposed [6]. However, tests with reacting substances are also possible, although then, for safety reasons, pressure relief will be effected at lower temperatures. On the plant scale, to ensure effective cooling of the contents by evaporation using a pressure relief system, it is important that the pressure relief be initiated as early as possible, since the mass flow of the evaporating product (usually solvents or monomers) necessary for the cooling is still small, foaming often does not yet occur, and the cross-section of the pressure relief device can remain small.

CONCLUDING REMARKS

I believe that my arguments demonstrate that effective control of thermal instabilities is only possible if the selected investigation methods are closely adjusted to suit the specific problem and the characteristics of the apparatus.

Standardization of investigation methods is not conceivable or at least not for problems on the production site. The presented examples and problems also illustrate that many tasks can only be resolved by a team comprising : physical-chemists, physicists and engineers, since detailed knowledge from all of these disciplines is almost always required.

REFERENCES

[1] J. O'Brien, et al
"Thermal Stability Hazards Analysis"
Chem. Eng. Prog. 78, p. 46 (Jan. 1982)

[2] H. Schacke, et al
7. Bunsen-Kolloquium "Thermische Stabilität und sichere Hand-
habung kondensierter Stoffe", Frankfurt-Hoechst 1982

[3] R. Gygax
"Assessment of potential risks due to runaway or unintended
reactions"
presented at the European Conference on Evaluation of Thermic
Hazards and Prevention of Runaway Reactions, Nov. 1982, Zürich,
Switzerland

[4] F. Brogli
"Heat Balance of Unstirred Containers"
Proc. 3rd Int. Symp. on Loss Prevention and Safety Promotion in the
Process Industry, 1980

[5] N. Schulz
"Beherrschung unerwünschter Reaktionsabläufe durch Druck-
entlastung"
Sichere Chemietechnik, Sicherheitstechnisches Kolloquium der
Bayer AG, 1982, p. 119

[6] L. Friedel, G. Löhr
Auslegung von Entlastungseinrichtungen für Gas-Flüssigkeits-
Reaktionssysteme, Verf. Techn. 15 (1981) Nr. 4, p. 259

RUNAWAY REACTIONS IN A POLYACRYLONITRILE WET-SPINNING PLANT

G. Arabito, V. Caprio, S. Crescitelli, G. Russo, V. Tufano

Istituto di Chimica Industriale e Impianti Chimici, Università
Istituto di Ricerche sulla Combustione, C.N.R.
P.le Tecchio, 80125 Napoli, Italia

ABSTRACT

The thermal instability of acrylic copolymer/nitric acid liquors was experimentally studied by means of a Sikarex III adiabatic calorimeter. Moreover, the evolved gases were analysed by gas-chromatography and different chemical analyses were performed on the liquors obtained by quenching the reaction after different reaction times. The measured concentrations of the reaction products were fairly well correlated by means of a mathematical model based on a four-step kinetic model, which includes the hydrolysis of the -CN groups and the decomposition of the intermediates.

INTRODUCTION

Runaway reactions can occur in the industrial plants for the wet-spinning of an acrylic copolymer from nitric liquors. In a previous paper /1/ it was shown that the reactivity of these liquors is very low (not detected by DSC); nevertheless, when they are reacted in adiabatic conditions, noticeable increases of temperature are recorded, large amounts of gases are evolved and the boiling of the nitric acid/water mixture is observed, so that dangerous overpressures and emission of toxic substances can occur.

The experimental data obtained by adiabatic calorimetry and the results of a simplified mathematical analysis performed with the usual methods /1,2/ indicated a rather complex reaction mechanism, but were inadequate to substantiate a quantitative kinetic model. On the other hand, such a model appeared necessary for the evaluation of the hazards related to the process and for the design of adequate safety devices. Therefore, a more detailed chemical study was initiated /3/, with the aim of measuring the concentration of the main intermediates and products. In this paper, the results of this study are presented and discussed, together with the calorimetric data, and a four-reaction kinetic model is proposed to describe the experimental results.

EXPERIMENTAL TECHNIQUES

All the experimental data reported here are relative to reaction tests performed by a Sikarex III adiabatic calorimeter /4/ on samples prepared by dissolving prefixed amounts of PAN (91.5% by weight polyacrylonitrile, 8% methylacrylate and 0.5% sodium metallulsulphonate)in azeotropic nitric acid/water solutions.

The noncondensable gases evolved during the experiments were collected at different times at the outlet of a water-cooled trap and analysed by gas-chromatography. The condensed vapors were analysed for HNO_3 by titration.

Several experimental runs were stopped after different, prefixed reaction times, to determine the composition of both the precipitated polymer and the washing waters. Potentiometric titrations were performed on the washing waters to measure the concentration of both nitric and carboxylic acids, whereas the ammonium concentration was evaluated by the Nessler method. The carboxylic acids were measured also in the precipitated polymer, by neutralization at 0°C with sodium hydroxide 0.2 N and back-titration of the excess reactant. Moreover, the precipitated polymer was analysed by IR spectroscopy to evidence the absorption-bands modifications and by elemental analysis to measure the nitrogen/carbon ratio.

Further details on the experimental techniques can be found in /3,5/.

EXPERIMENTAL RESULTS AND KINETIC MODEL

Most of the experimental results discussed here refer to 11% by weight PAN liquors, reacted in adiabatic conditions starting from an initial temperature of 30°C.

During the first stage of the reaction only a slight increase of temperature was measured and no gas evolution was detected. Subsequently, both the temperature and the gas flow rate increased sharply /1/. The comparison between the IR spectra of the unreacted PAN and that of a sample reacted up to this second stage (that is for about 22 hours), reported in figure 1, shows the disappearance of the $-CN$ peak at 2237 cm^{-1} /6/. Moreover the broadening of the peaks near 1700 cm^{-1} and the very broad absorption extending between 2300 and 3300 cm^{-1} can indicate the presence of carboxylic groups, whereas the strong absorption in the region 1500-1750 cm^{-1} together with the series of bands at frequencies higher than 3300 cm^{-1} suggests the presence of amidic groups.

Figure 2 reports the number of moles of carboxylic groups in the polymer and of ammonium ions in the washing waters, as a function of the reaction time. The maxima shown at times slightly greater than 20 hours and the following decrease indicate that both these species act as reaction

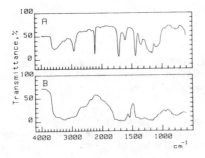

Fig. 1 - IR spectra of the acrylic copolymer PAN.
A) Unreacted PAN;
B) PAN reacted for about 22 hours in a 11% by weight liquor, starting from a initial temperature of 30°C.

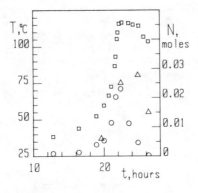

Fig. 2 - Sample temperature ($\square$), number of moles of ammonium ions (O) and of carboxylic groups (Δ) vs. the reaction time. 11% by weight PAN liquor (initial number of moles of -CN groups =0.052, initial temperature T_o =30°C.

intermediates.

These results, together with the observed constancy of the nitric acid concentration, allow to reject the hypothesis of nitration of the polymeric chain, and to suggest a two-step hydrolysis of the -CN groups, according to the scheme /7,8/:

$$-CN + H_2O \xrightarrow{\text{H}^+} -CONH_2 \qquad (1)$$

$$-CONH_2 + H_2O + H^+ \longrightarrow -COOH + NH_4^+ \qquad (2)$$

The figure 3 reports the amounts of noncondensable gases as a function of the reaction time. Nitrogen and carbon dioxide are by far the prevailing reaction products, whereas only small amounts of nitrous oxide, carbon monoxide and nitric oxide were detected. The total number of moles of

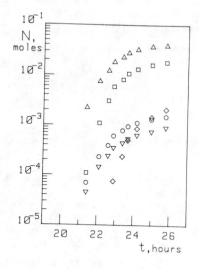

Fig. 3 - Number of moles of the
gaseous products vs. the reaction
time. 11% by weight PAN liquor
(initial number of moles of -CN
groups = 0.052)
$\triangle$ = nitrogen;
$\square$ = carbon dioxide;
$\bigcirc$ = carbon monoxide;
$\triangledown$ = nitrous oxide;
$\diamondsuit$ = nitric oxide.

nitrogen produced is equal to 0.8 times the initial number of moles of
-CN, whereas the number of moles of carbon dioxide produced is about half
the number of moles of -COOH reacted.

The stoichiometric ratios suggest the occurrence of two parallel reactions:

a) oxidation of the polymer by nitric acid, followed by an internal
esterification /9/, according to the scheme:

$$
\begin{array}{c}
\text{CH}_2 \\
\text{CH} \quad \text{CH} + \text{HNO}_3 \longrightarrow \text{CH} \quad \text{CH} + CO_2 + H_2O + HNO_2 \\
\text{COOH} \quad \text{COOH} \quad\quad\quad C - O
\end{array}
\tag{3}
$$

This reaction scheme agrees with the observed stoichiometry, because the
esterification is faster than the oxidation.

b) decomposition of the ammonium nitrate in acidic conditions according
to the scheme /10/:

$$5\ NH_4NO_3 \longrightarrow 4\ N_2 + 9\ H_2O + 2\ HNO_3 \tag{4}$$

The occurrence of this reaction is also confirmed by the results of
several runs performed with liquors to which prefixed amounts of ammonium
nitrate were added; also in this cases, at the completion of the reactions,
the molar ratio $N_2/(-CN + NH_4NO_3)$ was equal to 0.8.

The proposed reaction scheme can explain also the third reaction stage, during which the near-isothermal boiling of nitric acid/water mixtures was observed /1/, since all the four reactions (1)-(4) are exothermic.

No attempt was made to model the production of carbon monoxide, nitrous oxide and nitric oxide, because of the small amounts formed. Their presence indicates a more complex reaction scheme, that could probably explain also the decrease of temperature observed after the isothermal boiling at the lowest PAN concentrations /1/.

MATHEMATICAL MODEL

Reactions (1)-(4) were considered in the mathematical model, which is essentially formed by a set of five differential equations derived from the mass and the entalpy balances. The kinetic equations were assumed to be pseudo-first order, in the reactants $-CN$, $-CONH_2$, $-COOH$ and NH_4NO_3 respectively; this assumption is justified by the large excess of water and nitric acid in the system.

The thermodynamic data were taken from the literature /11/, whereas the heat capacity of the polymer was measured by the Sikarex calorimeter. The boiling temperature was assumed equal to the experimental value; this temperature is only few degrees lower than that computed for binary nitric acid/water solutions.

The values of the kinetic parameters were estimated by minimizing the deviations between the computed and the measured values. The computed results are compared with the experimental data in the figure 4, for a 11% by weight PAN liquor, reacted at an initial temperature of 30°C. A fairly good agreement is observed. The values of the kinetic parameters obtained

TABLE 1

Entalpies of reaction evaluated from the literature data /11/, and kinetic parameters for pseudo-first order reactions. The pre-exponential factors refer to 11% by weight PAN liquors.

Reaction n.	Entalpy of reaction (kcal/mole)	Pre-exponential factor (hr^{-1})	Activation energy (kcal/mole)
1	21	3.75 E 11	19
2	10.4	5.0 E 10	18
3	19.8	2.7 E 8	16
4	58.8	1.3 E 6	11

with this procedure are reported in Table 1; the activation energies of the two hydrolysis reactions are in good agreement with the values reported by Kundryavtsev and Zharkova /8/.

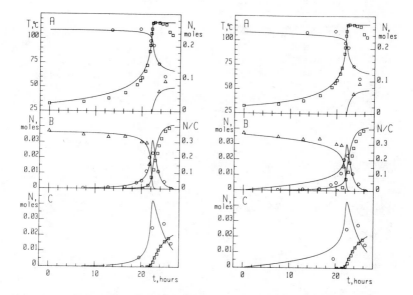

Figs. 4 and 5 - Experimental data (11% by weight PAN liquors, initial
number of moles of -CN groups = 0.052) and computed data (fig. 4, four-
reactions model, fig. 5, three-reactions model).
A) □ = temperature; O = nitric acid in the system;
   △ = nitric acid condensed in the water-cooled trap;
B) △ = nitrogen/carbon ratio; O = ammonium ions; □ = nitrogen;
C) O = carboxylic groups; □ = carbon dioxide.

On the contrary, the modeling of the hydrolysis process by a single
overall reaction (assuming the reaction (1) much faster than reaction (2)
/8/):

$$-CN + 2 H_2O + H^+ \longrightarrow -COOH + NH_4^+ \tag{5}$$

leads to overestimating the ammonium concentration and to underestimating
the nitrogen/carbon ratio, as shown in figure 5.

In figures 6 and 7 the computed results are compared with the experimental
data, for different initial temperatures at constant PAN percentage (11%
by weight) and for different initial PAN concentrations at constant
initial temperature (T_o=30°C) respectively. Also in this case, a fair
agreement is observed; in particular, as far as the last stage
(subsequent the isothermal boiling) is concerned, the mathematical model
simulates with sufficient accuracy the behaviour observed at the highest
PAN concentration (figure 7) whereas at the lowest PAN concentrations
(figures 6 and 7) an isothermal profile is computed, because reactions

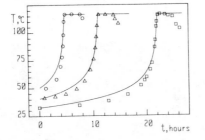

Fig. 6 – Experimental and computed data (four-reactions model) for 11% by weight PAN liquors at different initial temperatures.
O = 50°C
△ = 40°C
□ = 30°C

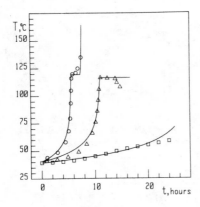

Fig. 7 – Experimental and computed data (four reactions model) for different PAN percentages.
O = 20% by weight
△ = 11%
□ = 5%

and boiling are complete and the system is adiabatic. The decrease of temperature observed experimentally cannot be modeled because no endothermic reaction is included in the reaction scheme.

CONCLUSIONS

The temperature vs. time data measured by an adiabatic calorimeter and the results of different chemical analyses on the reaction products allowed to single out the main reactions that produce the thermal instability of acrylic copolymer/nitric acid liquors. A mathematical model was derived, which simulates with enough accuracy the runaway reactions. Therefore the model presented can be used to extrapolate the calorimetric data to actual operative conditions of industrial plants.

It may be worth noting that these results could not have been achieved by the use of the calorimetric data alone.

REFERENCES

/1/ V. Caprio, S. Crescitelli, V. Piccolo, G. Russo, V. Tufano,
Journal of Hazardous Materials, in press (1983)

/2/ D.I. Townsend, J.C. Tow, Thermochim. Acta, 37, 1 (1980)

/3/ G. Arabito, V. Caprio, S. Crescitelli, G. Russo, V. Tufano,
Proceedings of the 4° Convegno nazionale di calorimetria ed analisi
termica, 107, Catania, (1982)

/4/ V. Caprio, S. Crescitelli, G. Russo, V. Tufano, Convegno su "Strumen-
ti e metodi per la valutazione della sicurezza chimica", Milano (1982)

/5/ G. Arabito, V. Caprio, S. Crescitelli, G. Russo, V. Tufano,
Proceedings of the Convegno CISACH 1°, Milano (1983)

/6/ J. Brandrup, E.H. Jmmergut, Polymer Handbook, J. Wiley and sons,
New York p. VI-71 (1966)

/7/ J. Hine, Physical Organic Chemistry, Mc Graw Hill, New York (1962)

/8/ G.I. Kudryavtsev, M.A. Zharkova, J. of Applied Chemistry (Engl.
Translation), 29, 1189 (1956)

/9/ T. Matsmumoto, Chem. High Polymers 7, 142 (1950)

/10/ L. Medard, Les Explosifs Occasionnels, Technique et Documentation,
Paris (1979)

/11/ Landolt-Börnstein, Zahlenwerte und Funktionen, Vol. IV, part 2,
Springer, Berlin (1976)

Thermohydraulic processes in pressure vessel and discharge line during emergency relieving.

Lutz Friedel, Hoechst AG, D 6230 Frankfurt/M. 80
Stefan Purps, Dechema, D 6000 Frankfurt/M. 97

ABSTRACT

Systematic vessel depressurisation experiments in a laboratory and on a pilot plant scale with initially saturated refrigerant R12 as model fluid show the influence of initial liquid level, relief cross section, starting pressure, relief line geometry and vessel volume on boiling or liquid level swell delay, thermodynamic disequilibrium, vapour generation, onset and duration of two phase flow, total relief time etc.. Additional temperature, two phase mass flow and void fraction measurements indicate a weak thermodynamic disequilibrium and temporary occurrence of a pulsating semi-annular flow in the discharge line.

INTRODUCTION

The complex instationary thermodynamic and fluiddynamic processes associated with emergency top venting of reaction vessels with gas/vapour-liquid mixtures initiated by an uncontrolled pressure increase caused by a runaway reaction, a heat incident as a result of fire hazard or damage to the vacuum insulation of a liquid or liquified gas storage tank or simply by human or technical failure are still not completely understood.

Basic depressurisation experiments with initially saturated refrigerant R12 as model fluid are therefore carried out to investigate - for the present without the implications of a superposed reaction - the transient pressure and temperature response in a conventional top vented, non stirred pressure vessel and in the discharge line. Additional parallel measurements of mean void fraction and liquid or mixture level swell in the vessel and of mass flow and void fraction in the relief line aimed at exploring the hydraulic interactions. Parameters of the systematic experiments are initial liquid level, depressurisation rate (relief cross section), starting pressure, relief line diameter and vessel volume, as all these variables are indispensible in practical application as intermediate states, in safety assessments and for scaling purposes.

The results of this still initial investigation are considered as a basis for studying the instationary thermohydraulic behaviour of real and more viscous chemical mixtures and the complex hydraulic interactions between vessel and relief line, in order thereby to develop a generally applicable blowdown computer code for design purposes and safety analyses of operational and accidental transients.

TEST FACILITY

The test facility in a laboratory and on a pilot-plant scale consists of pressure vessel (15 or 107 l), discharge line (DN 20 or 50) and receiver vessel (8 m^3), Fig. 1. The instrumentation includes conventional shielded thermocouples and pressure transducers in the vessel and pipings. Furthermore, a two phase mass flow meter is inserted into the discharge line and a capacitance liquid or mixture level swell and void gauge is located in the pressure vessel, Fig. 2. All primary measurements are logged by a dedicated Kontron KAP 1000 computer system.

The depressurisation process is initiated by opening of a fast acting ball valve and controlled by means of orifices with different bores alternatively inserted in the discharge line. A full description of the apparatus and the instrumentation is given elsewhere /1/.

EXPERIMENTAL RESULTS

Pressure traces in the vessel

Opening of the fast acting valve in the relief line is followed - as already described elsewhere (e.g. Scriven /2/) - by a characteristic sudden decrease in vessel system pressure due to vapour outflow, a subsequent pressure recovery (only pronounced at high depressurisation rates) during the vapour generation stage leading to mixture level swell, and then finally by a moderate and uniform fall of pressure towards the (back) pressure in the receiver vessel, Fig. 3. The trace of the saturation pressure developed from the measured liquid bulk temperature indicates in comparison to the system pressure that initially the mean liquid temperature in the vessel remains to a fairly large extent unchanged during the pressure undershot process, resulting in a substantial liquid superheat and thermodynamic disequilibrium between liquid an vapour respectively. Only after the pressure recovery stage does it gradually approach the vessel pressure. Indeed, in the near wall and near bottom region as well as at the free liquid/vapour interface, the temperature and the corresponding local saturation pressure response respectively are faster and drop far earlier depending on the depressurisation velocity.

With high relieving rates, an extreme and, due to the negligible hydrostatic pressure difference across the liquid height, uniform thermodynamic disequilibrium between liquid and vapour phase is instantaneously built up and a heterogeneous nucleation and a sudden vapour generation at the (wetted) vessel wall and bottom crevices is initiated, whilst homogeneous nucleation and surface boiling at the free interface - evoked by backfalling drops of (depressurised) condensed vapour acting as nuclei - occur to a smaller extent. The boiling front caused by the heterogeneous nucleation advances towards the top of the vessel, whereby the difference in time between the change of temperature in the near bottom and in the bulk region roughly corresponds to the free rising time of bubbles with a velocity of 0.4 m/s. The time delay, however, vanishes with decreasing initial liquid level, as then only moderate superheating arises in the liquid. The surface boiling is limited to the interface region and does not propagate far downwards. This should only occur with very pure and degassed fluids in glass tanks with smooth crevice-free walls /3/.

With low relieving rates the evaporization here preferentially sets in at the interface through surface boiling and the heterogeneous nucleation is to a large extent suppressed, since the vapour production at the interface already suffices to reduce the superheating or disequilibrium.

On comparing the intensity of vapour production, mixture surface displacement and liquid entrainment into the relief line, it becomes clear that in cases of prevailing heterogeneous nucleation a higher liquid holdup in the flow and a longer duration of the two phase flow is maintained.

It is now shown that the characteristic pressure trace changes with variation of initial liquid level in the vessel, Fig. 4. For all four liquid level ratios there is initially a practically linear decrease in pressure with increasing time or proceeding depressurisation. The different gradients result from the respective magnitude of the vapour volume above the liquid in the dome: With the highest initial liquid level the vapour volume is smallest and since in all cases initially the same vapour volume flux flows out, depressurisation is at its greatest. At the same time the most extreme disequilibrium and hence the highest superheating of the liquid is attained, resulting in the most explosive vapour generation and pressure recovery. With low liquid levels the generation of vapour occurs more smoothly and in extreme cases no distinct pressure undershot will be notified. As expected, total depressurisation of the vessel with a higher liquid level lasts longer.

These typical pressure traces are only obtainable, if the saturated two phase system is briefly relieved several times before starting the experiments. Some pressure curves without initial degassing are presented in Fig. 5. They are also readily reproducible and the question now still remains as to which experimental technique or liquid is representative for industrial applications.

The influence of the depressurisation rate on the pressure history caused by different control orifices in the relief line is shown in Fig. 6. At the beginning of all five experiments the same vapour volume is present in the dome above the liquid. As expected, during the initial time up to the pressure recovery stage the maximum vapour outflow and hence pressure decrease is obtained via the largest relief opening area. At the same time the highest disequilibrium or liquid superheating again occurs in this situation and the vapour generation starts earlier and is more active. Pressure recovery is therefore very pronounced. With the smallest control orifice, the pressure decreases smoothly and almost uniformly and during the initial stage only a negligible disequilibrium occurs. It is worth pointing out that due to the pressure recovery after the initial depressurisation stage a practically identical intermediate pressure is reached. Obviously, for practical applications it is advisable firstly to relieve via a small device and later on by means of an additional one.

An increase in the starting pressure with identical initial liquid level etc. is coupled with a greater vapour mass above the almost unchanged liquid mass and a higher critical mass velocity. The expansion of the vapour volume in the dome at the beginning of the relieving process therefore proceeds in all cases practically with the same pressure gradient. Due to the lower heat of vaporization at higher pressures the onset of boiling sets in earlier. Consequently, the pressure undershot is weaker and by way of suggestion may be reduced under extreme conditions to a form of saddle point in the pressure trace. Furthermore, only fairly small liquid superheating occurs. For that reason the period of time with mixture surface swell and displacement to the outlet and of two phase flow in the relief line does not last as long, leading to a relatively higher initial pressure reduction. But generally speaking, the relief time is longer because of the higher starting pressure and energy state, Fig. 7.

A change in the relief line diameter whilst retaining identical orifices and lengths produces some differences in the initial depressurisation stage: An increase in diameter means a lower resistance for the outflow and should therefore result in a quicker emptying of the vessel. On the other hand a greater additional vapour volume is entrapped between vessel flange and fast acting valve. When starting the depressurisation process this greater vapour mass must first flow out, since only than can the pressure in the vessel decrease. In the present experiments, the entrapped vapour volume increased from 3 to 25 % of the total vapour and hence caused a time delay of nearly 60 ms before a noticeable depressurisation could take place, Fig. 8. Overall, the depressurisation via the relief line with greater diameter is faster, since the effect of the lower flow resistance is preponderant. Indeed, the orifice in this situation exhibits not the essential pressure loss in the line, since with identical orifice cross sections different relief times apply.

For scaling purposes, the influence of the (total) volume of geometrically and materially similar vessels (equal H/D ratios) on depressurisation is compared in Fig. 9a, 9b. On maintaining in both experiments an identical ratio between vessel volume and relief cross section and an equal initial liquid level ratio, it immediately follows from single phase hydraulics that with identical starting pressure the depressurisation rates at the beginning of the relief process must coincide if, in addition, equal flow contraction coefficients are assumed. However, systematic differences appear to exist with the inception of boiling and the following relieving process. Firstly, the pressure recovery is more pronounced in the case of the smaller vessel, since due to its more advantageous relationship between liquid volume and wetted wall surface it offers a relatively higher number of activable crevices for (heterogeneous) nucleation. On proceeding with the pressure curve the depressurisation rate of the smaller vessel is for a certain period greater. Later on, this tendency changes and the pressure decay in the small vessel is slower. This results from the specifically higher heat transfer from the surroundings to the vessel cooled down by the evaporation and vapour expansion. Shortly before both vessels are totally relieved again the pressure traces intersect. However, this is only an experimental effect: in the test with the larger vessel a higher back pressure in the receiver vessel is built up due to the greater overflowing mass. Summing up, with increasing vessel volume the depressurisation by atmospheric venting never lasts longer if equal starting conditions prevail.

The transient pressure traces in the vessel during the relieving process are to a large extent intimately connected with the developing time-dependent phase distribution. In Fig. 10a, 10b a series of typical traces of the void fraction measured at different times after initiation of depressurisation in seven vertical positions - equidistant over the vessel height - is presented. Initially, the capacitance gauges C1 to C5 dip into liquid while with gauges C6 and C7 pure vapour is measured. After

250 milliseconds a substantial amount of vapour is present in the inter-
phase and bottom region. At the same time the mixture level has increased
and a two phase mixture of vapour and some droplets rises through the
vapour dome to the outlet. At the next stage considered (500 ms), a downward and an upward expansion of the two separated vapour zones can be
observed. Furthermore, the mixture level and the void fraction in the mixture as a whole increased. After 750 milliseconds the two zones have grown
together and due to the intensive agitation of the mixture caused by the
vapour production occuring now all over the whole superheated fluid volume
a practically uniform void distribution as a function of height is
obtained. During the next two recorded dates the mixture level extends up
to the top of the vessel and a linear increase of void fraction with the
altitude prevails. Indeed, in the latter sheet it is more pronounced due
to the imperfect phase separation and definite residence time of the
rising vapour phase in the mixture. After this stage, a characteristic
phase distribution with three distinct zones is found for the next three
measuring periods. It starts with a linear increase of the void fraction
in the near bottom region, followed by an intermediate zone with constant
void fraction which changes to a district with exponentially increasing
vapour fraction. In the lower and upper region an unrestrained separation
seems to occur while in between due to intensive momentum exchange and
flow pattern effects the rising of the vapour phase is still somewhat
restricted. Only now, between the 7th and 12th seconds is there a
suggestion that the mixture level is starting to drop. Later on (21 and 42
seconds), the zone with a constant void fraction has disappeared indicating complete separation of the vapour from the slightly superheated
mixture. Simultaneously, the mixture level is further dropping, after 42
seconds an equalisation of the pressure in the test and receiver vessel as
well as thermodynamic equilibrium is attained.

In addition to the overall response of the system, the point of time for
the occurrence of two phase flow in the relief line is also of interest.
Basically, a two phase mixture is only formed in the vessel when the
boiling delay time has elapsed and sufficient vapour has been produced so
that the swelling mixture level can reach and block up the outlet flange.
A prerequisite for this is (for R12) an initial depressurisation velocity
of more than 7 bar/s and an initial liquid level ratio exceeding 60 %. Interpretation of the first void fraction and mass flux measurements and simultaneously recorded pressure and temperature curves in the relief line
during the whole depressurisation process shows that fully developed two
phase flow can be expected as early as 600 - 800 ms after release of the
blowdown. A typical trace of the fluid temperature and of the saturation
temperature developed from the pressure in the relief line just behind
the orifice is depicted in Fig. 11. Four characteristic stages or flow
patterns can be distinguished: During the first (0 - 200 ms) the
(calculated) saturation temperature slopes down in accordance with the
rapid drop of pressure in the relief line and only saturated vapour flows
out. The fluid temperature follows immediately and in respect of its magnitude practically agrees with the former due to drops of condensed vapour
carried along in the saturated vapour flow which continuously comes upon
the thermocouple and ensures for the identity of the temperatures.

In the course of the following period lasting roughly 400 - 600 ms, both
temperatures diverge in some cases up to 15 K. The saturation temperature
further decreases due to the progressive drop in system pressure in the

relief line, but the fluid temperature impulsively rises as a consequence of the outflow of superheated vapour originating from the vapour generation now taking place in the vessel and as early as the thermocouple is wetted by first liquid (drops) again steeply breaks down, arriving at the saturation temperature and thus indicating the onset of two phase flow. This is further confirmed by void measurements and is accompanied by only a small thermodynamic disequilibrium in the line.

During the next stage, the mass flow quality of the two phase flow gradually increases. Furthermore, both temperatures firstly rise in line due to the pressure recovery in the vessel which advances through the relief line and then uniformly slope down, whereby the course of the saturation temperature for most of the time is slightly below. However, during the latter section of this (third) period the two temperature traces intersect and the arrangement is changed. But this is attributed to a measuring rather than to a physical effect.

The abrupt increase in the fluid temperature away from the uniformly decreasing saturation temperature marks the beginning of the fourth period of the depressuration process and the change of the two phase flow to a superheated single phase vapour flow. A physical explanation of this sudden temperature excursion in the vapour flow is not yet available but it is assumed that the superheat originates from the heating of the vapour in the vessel dome by the warm (vessel) walls. A similar effect in vessels during the depressuration process is only later observed due to the wetting of the thermocouple by entrained droplets and reported elsewhere /1,5/. As a result, the vapour phase is also not in thermodynamic equilibrium with the state of the remaining mixture in the vessel, its temperature after an isentropic expansion to the local pipe pressure would correspond to the measured superheated vapour flow temperatures in the line within the same order of magnitude.

Summing up, it is again clearly demonstrated by the three traces that the starting pressure affects the extent of the two phase flow period and hence of the total depressurisation process and relieving time. The explanation is already given in the interpretation of the traces in Fig. 7.

The flow patterns prevailing in the relief line well behind the orifice are developed from the records of the capacitance void fraction gauge. Its design permits the determination of the mean void fraction and simultaneously of the local values within the outer and inner annular gap with equal capacitance. Indeed, it is supposed that this cylindrical configuration only negligibly affects the development of the phase distribution in the flow.

Just after the start of the transient relief process, in both annulli a void fraction value of 0.97 - 0.98 is measured, indicating a flow of uniform wet vapour, Fig. 12. The local void fractions then decrease and an unsteady two phase flow takes place. It always lasts for a short time in the range of milliseconds, breaks down and again grows up. However, with increasing relief time the frequency of occurrence increases and gradually the flow pattern again changes to spray flow followed by single phase vapour flow characterised by equal void fraction values near unity.

During these unsteady two phase flow periods the distribution of the phases across the flow area in fact differs somewhat from the convential assumption of homogeneous flow. In the outer annulus, especially during the initial period, an appreciably higher liquid content is measured, leading to the ascertainment of semi-annular-mist flow consisting of a liquid film along the pipe wall and droplets in the core. Later on, there is the already mentioned change to vapour flow with entrained droplets in the core.

With increasing relief cross section the onset of two phase flow occurs earlier and the liquid content in the flow is higher. It can be accounted for by the more extreme thermodynamic nonequilibrium and extensive level swell due to the faster depressurisation. Additionally, the pulsations are more powerful and overall the two phase flow period continues longer.

The transient course of the mass flow rate in two hand picked experiments with different depressurisation rates is shown in Fig. 13. After a short take-off time of about 40 milliseconds a critical outflow of vapour occurs and as expected via the greater orifice area a higher outflow of mass is obtained. With the occurrence of a two phase flow the mean velocity of the mixture substantially decreases and the total mass flow drops, too. However, the appearence of the first liquid slug announcing two phase flow (something more than vapour/droplet flow) is first of all coupled with a very short-ranged increase of the mass flow. Within the next 500 milliseconds the (two phase) mass flow is then lowered to about half of that of the critical vapour flow whereby the volumetric quality of the liquid only amounts between 4 and 20 %. In the following, with increasing void fraction and relieving time the mass flow again gradually rises towards higher values and finally arrives at a critical vapour mass flow. Indeed, this description holds only for tests with small orifice areas if simultaneously a high liquid holdup prevails in the flow.

CONCLUSION

The transient data obtained so far reveal the complex interaction between the process in the vessel and in the relief line during depressurisation and the influence of the macroscopic variables of the experiments such as initial liquid level, relief cross section, starting pressure, relief line geometry and vessel volume. The results agree with the recommended practice to avoid two phase flow as the the total mass flow and cooling capacity decreases and firstly to relieve via a small safety device and later on by means of an additional one. Finally, it can be concluded that also for rough estimations it is not allowed to calculate only with the vapour volume in the dome instead of applying the complete energy and mass conservation laws to the system. In this case the assessment of the duration of the relieving process falls short by a factor of 5 to 10.

LITERATURE

/1/ L. Friedel, S. Purps: Chem.-Ing.-Tech. 54 (1982) 3, 256/257.
/2/ A.H. Scriven, P.R. Farmer: Paper presented at the European Two Phase
                          Flow Group Meeting, Glasgow, 1980.
/3/ H.-J. Viecenz: Diss. Univ. Hannover, 1980.
/4/ N. Schulz: Diss. RWTH Aachen, 1975.
/5/ D.W. Sallet et al.: Proc. Inst. Mech. Engrs. 194 (1980) 26, 225/230.

ACKNOWLEDGEMENT

The financial support provided by the (German) Bundesminister für
Forschung und Technologie is gratefully acknowledged.

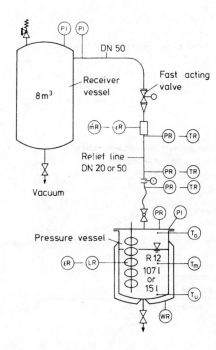

Fig. 1: Test facility

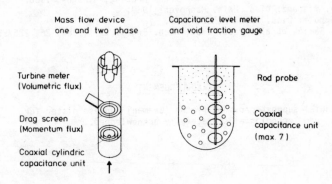

Fig. 2: Single and two phase flow instrumentation

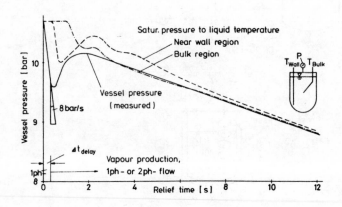

Fig. 3: Characteristic pressure trace in the vessel

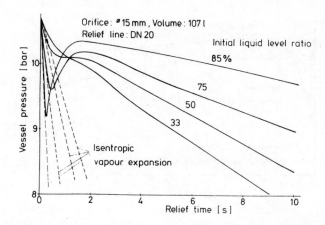

Fig. 4: Pressure traces at different initial liquid levels

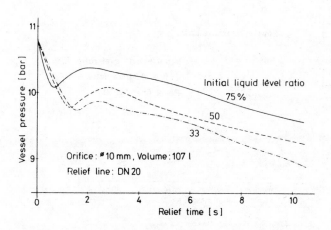

Fig. 5: Pressure traces without degassing before the experiments

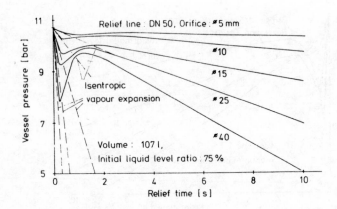

Fig. 6: Pressure traces at different orifice diameters

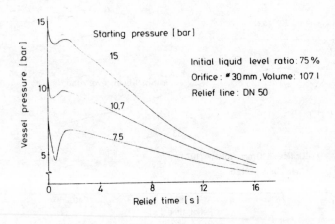

Fig. 7: Pressure traces at different starting pressures

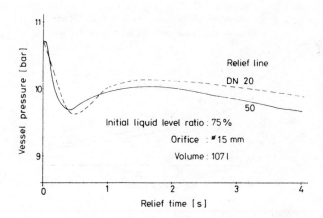

Fig. 8: Pressure traces at different relief line diameters

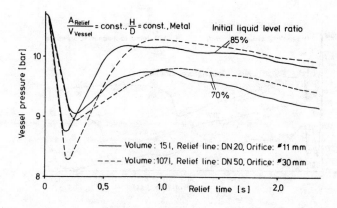

Fig. 9a: Pressure traces at different vessel volumes, initial stage

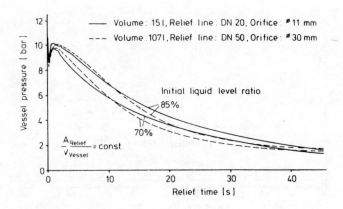

Fig. 9b: Pressure traces at different vessel volumes

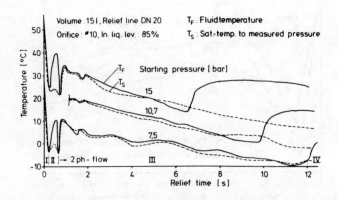

Fig. 11: Temperature traces in the relief line behind the orifice

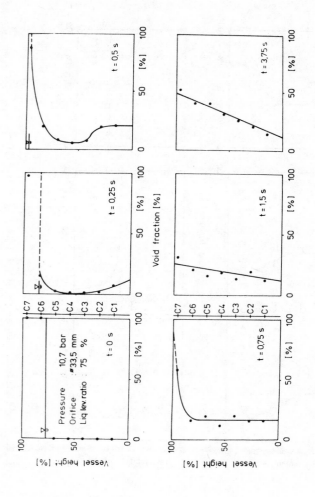

Fig. 10a: Void distribution at different times after initiation

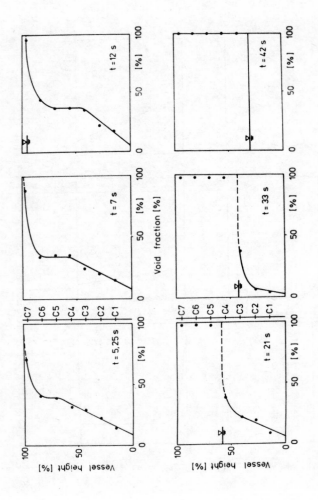

Fig. 10b: Void distribution at different times after initiation

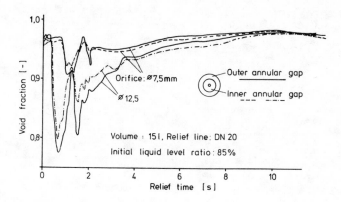

Fig. 12: Void fraction in the relief line behind the orifice

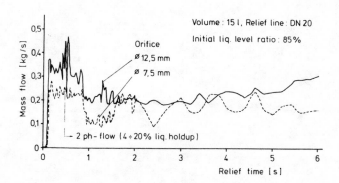

Fig. 13: Mass flow in the relief line behind the orifice

Unstable Products

Chairman

T. Grewer
Hoechst, Germany

SELF-ACCELERATING DECOMPOSITION OF ETHYLENE CYANIDE

W. BERTHOLD, U. LÖFFLER

BASF AG, Ludwigshafen, Fed. Rep. Germany

Introduction

A self-accelerating reaction is characterized by an increasing reaction rate at constant temperature as opposed to a "normal" reaction with only temperature-controlled reaction rates. Self-accelerating reactions are potentially more dangerous than temperature-controlled reactions since the reaction may occur suddenly and unexpectedly. Its progress within a given time may be influenced by its thermal history as well as by unknown external influences.

When applying routine test methods such as Differential Thermal Analysis (DTA/DSC) or adiabatic storage the self-accelerating mechanism of a reaction may be not detected. However, there are some indications which should be noticed: peak temperatures evidently influenced by different DTA scan speeds, or adiabatic runaway reactions carried out at much lower temperatures than the onset temperatures in DTA experiments.

A reliable method of detecting a self-accelerating reaction is isothermal DTA[1]. An equivalent method, which will not be described in detail, is a DTA in which the temperature is increased stepwise[2] or a DTA[3] using a sample which has been stored at a temperature below the onset temperature of a standard DTA. The self-accelerating behaviour may then be described by the isothermal induction period as a function of temperature.

As will be shown, in the case of ethylene cyanide the decomposition occurs suddenly without heat being evolved during the induction period. For this reason we found the same induction periods in adiabatic storage tests as in isothermal experiments.

Results

a) Measurements on the pure product

Fig. 1 shows the result of a standard DTA of pure ethylene cyanide.

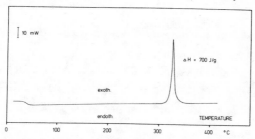

FIG. 1

DECOMPOSITION OF ETHYLENE CYANIDE

DTA   sample : 22,5 mg .   sample holder :   Hastelloy .   scan speed :   2 K / mn

The onset temperature is about 300 °C.
Integration of the very narrow peak (suggesting self-accelerating behaviour of the substance) leads to a total heat of about 700 J/g.

Fig. 2 shows the result of an isothermal DTA carried out 90 °C below
the onset temperature.

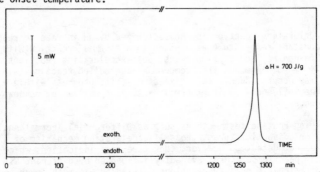

FIG. 2

DECOMPOSITION OF ETHYLENE CYANIDE

Isothermal DTA : T = 210 °C

After an induction period of more than 20 hours the heat evolution
sets in suddenly. The peak area is equivalent to 700 J/g equal to the
heat evolution derived from standard DTA. This means that no heat was
evolved during the induction period.

In Fig. 3 the isothermal induction periods τ for different temperatures
T have been plotted.

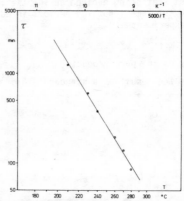

FIG. 3

DECOMPOSITION OF ETHYLENE CYANIDE

Induction period τ as a function of starting
temperature T from isothermal DTA

The function $\ln \tau \sim 1/T$ is the same as for adiabatic induction periods of "normal" (i.e. only temperature-controlled) reactions.

If the sample has been kept at a particular temperature for a certain period and then cooled down, the induction period at the same temperature will be shortened by a similar period. This means that the product "memorizes" its thermal history.

b) Reaction mechanism

A formal kinetic model was developed[4] based on the mechanism of radical chains, which is able to describe all isothermal DTA measurements.

The latest results indicate however, that the following reaction mechanism is more probable:

$$CN - CH_2 - CH_2 - CN \xrightarrow{CN^-} HCN + CH_2 = CH - CN$$

At elevated temperatures ethylene cyanide decomposes autocatalytically, accelerated by CN^-, into hydrogen cyanide and acrylonitrile. The sudden evolution of heat is due to the spontaneous polymerisation of hydrogen cyanide and acrylonitrile.

As shown in Fig. 4 and 5, both decomposition products polymerize by a self-accelerating mechanism, hydrogen cyanide reacting similarly to ethylene cyanide.

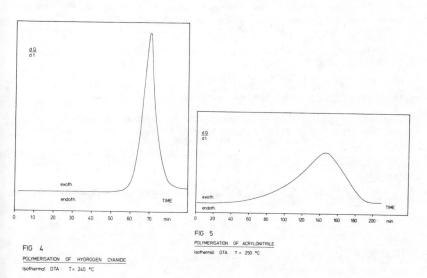

FIG 4

POLYMERISATION OF HYDROGEN CYANIDE

Isothermal DTA   T = 240 °C

FIG 5

POLYMERISATION OF ACRYLONITRILE

Isothermal DTA   T = 250 °C

Because of the complexity of the kinetic model many more fit parameters would be available to describe the experiments. However, we have decided against developing a new mathematical model as it would not corroborate the proposed decomposition theory.

c) Influence of additives

To anticipate the result: In experiments using more than 2 dozen different additives not a single stabilizing substance for ethylene cyanide was found. With conventional radical acceptors no effect was observed, whereas acids and bases lowered the thermal stability of ethylene cyanide.

By the addition of 1 % by weight HCN the induction period was reduced to 20 % of the induction period for the pure substance.
By 1 % (weight) KCN the induction period was dramatically reduced by a factor of 50, which is an indication of the correctness of the model.

Conclusions

It seems to us that there are many more self-accelerating reactions than have been published. This lack of knowledge may be explained by the fact that this potentially dangerous property may not be detected by routine measurements such as standard DTA or adiabatic storage tests. Therefore many substances or reaction mixtures are believed to be stable at a certain temperature and are handled without suspicion for years until an unexpected thermal explosion occurs, e.g. after the product has been held at a high temperature longer than usual.

Since isothermal DTA helps detect self-accelerating reactions inexpensively, this method should be applied as a matter of routine. On the other hand, it will not usually be possible to detect the reaction mechanism. As shown for ethylene cyanide, many experiments have had to be run, which is not practicable for the large numer of substances in the chemical process industries. However the chemist should be aware of the self-accelerating mechanism and he should be given the following recommendations:

- as short residence times as possible

- avoid impurities or additives with unknown effects

The maximum residence time as a function of temperature should not only be derived from isothermal DTA measurements, because in many cases self-acceleration and thermal acceleration overlap.

References

[1]F. Brogli, P. Grimm, M. Meyer, H. Zubler, 3rd International Symposium, Basle/Switzerland, September 15-19, 1980, Volume 2.

[2]H. Dörr, 6th international conference on thermal analysis, Bayreuth, federal republic of germany, july 6-12, 1980, conference work book 244 HL

[3]G. Hentze, Vorträge des Rapperswilers TA-Symposiums, April 18-20, 1979, experientia supplementum 37

[4]G. Eigenberger: Ein kinetisches Modell für die Zersetzung von Bernsteinsäuredinitril, unpublished.

THERMAL EXPLOSION OF LIQUIDS

J.VERHOEFF

PRINS MAURITS LABORATORIUM TNO, P.O.BOX 45, 2800 AA RIJSWIJK

1. Introduction

At present an explicit need exists to characterize (or classify) hazardous substances and reaction systems in order to enhance the safety of production, transport, storage and handling of chemical compounds. Although much attention has been given to thermal hazards, explosion phenomena resulting from a thermal runaway of liquids have hardly been investigated. The intensity of such a thermal explosion is of great consequence for the safety provisions to be incorporated in a process design. Without a thorough knowledge of the fundamentals of such an explosion it is virtually impossible to distinguish between the hazards of different reaction systems because in many cases explosions only occur after a thermal runaway.

A thermal explosion of a liquid appears to be rather complex and can be divided into three subsequent stages (1):

- the thermal runaway stage,
- the initiation stage i.e. a stage in which a hot kernel arises and
- the explosion stage, i.e. explosion in the proper sense.

This paper describes a few results of thermal explosions of a model substance - tertiary butylperoxybenzoate (TBPB) - in a low pressure autoclave. Kinetic data from the thermal runaway stage are compared with literature data.
Only a few main features of the thermal explosion of TBPB, which are also of importance for other substances or reaction systems, are treated.

2. Low pressure autoclave

The capacity of the autoclave sketched in Figure 1 is 0,01 m³. The autoclave consists mainly of a heated sample compartment and an isothermal compartment in which the explosion products expend. The maximum working pressure amounts to 4,0 MPa at a temperature of 473 K. The sample compartment has been designed in such a way that explosions caused by self-heating can be investigated. A gas-space of 1 mm between the inner sample vessel and the autoclave takes care of the required heat resistance to provide the conditions for a thermal runaway after heating the liquid sample. A 4-blade turbine stirrer has been mounted to diminish temperature differences in the sample mass. Temperatures in the sample vessel are measured by means of two chromel-alumel thermocouples with a response time of 20 ms in liquid. Simultaneously the pressure is measured with a piezoresistive transducer. Temperature and pressure signals are registered continuously on an instrumentation recorder and may be digitalized for further computer analysis.

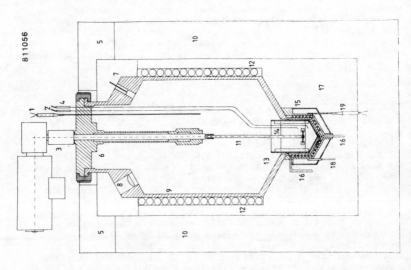

Figure 1. Low pressure autoclave.

1. Thermocouple autoclave
2. Thermocouple sample vessel
3. Stirring motor and transmission
4. Thermocouple sample vessel
5. Insulation cover filled with polyurethane
6. Cover of the autoclave
7. Pressure transducer
8. Safety membrane
9. Autoclave
10. Insulation jacket filled with polyurethane
11. Turbostirrer
12. Cooling coil (thermostat)
13. Sample vessel
14. Baffles
15. Cooling jacket (cryostat)
16. Electric heating
17. Space around pressure vessel filled with
    insulation material
18. Platinum thermometer
19. Thermocouple

811056

3. Thermal runaway stage of tertiary butylperoxybenzoate (TBPB)

Tertiary butylperoxybenzoate (TBPB) is widely used as a source of free radicals for polymerization and belongs to the group of peroxyesters which may decompose via a single or multiple bond-breaking process (2,3,4).
An extensive kinetic study of the stability of TBPB and para-substituted peroxybenzoates has been performed by means of differential scanning calorimetry (5,6,7). The experimental rate equation for TBPB determined for diluted (0,1 M) solutions in the range 373 K - 413 K is:

$$\ln k = \ln (2,1.10^{10}T) - \Delta G*/RT \tag{1}$$

where k is a first order rate constant, T is the absolute temperature, $\Delta G*$ is the free enthalpy of activation = 127,7 ± 0,6 kJ mol^{-1} and R is the gas constant.
If the Arrhenius relation is applied to describe these DSC results the first order rate equation is:

$$\ln k = \ln k_o - E/RT \tag{2}$$

where $\ln k_o$ = 33,3 ± 0,4 and E is the activation energy = 139,0 ± 1,7 kJ mol^{-1}.

Solution of the governing heat and mass balance equations during the thermal runaway stage of a low pressure autoclave experiment (Figure 2) results in reaction kinetic data. Investigation of technical pure TBPB revealed that the following heat generation formula holds:

$$q_m = 7,4.10^{19} \exp(- 128000/8,314\ T) \tag{3}$$

    temperature range: 380 K - 430 K
    pressure range   : 0,1 MPa - 0,15 MPa
    conversion range : 0 - 0,12

where q_m is the heat generation per unit of initial mass.
The respective standard deviations for the heat generation factor and the activation energy are 0,10.10^{19} W kg^{-1} and 2,0 kJ mol^{-1}. The heat of explosion of TBPB amounts to 1380 kJ kg^{-1}.

Compared with the above-mentioned literature data there are two striking differences: firstly, the upper value of the temperature range covered by the low pressure autoclave is higher than that of DSC (5,6,7) in spite of the large difference in concentration, and secondly the zero order equation (3) to describe the autoclave results as compared to the first order description of the DSC results (equation (2)).
The first point elucidates that with the low pressure autoclave reaction kinetic data are determined under thermal runaway conditions thus restricting the extrapolation of kinetic data to a practical situation to a minimum. The second point is connected with the phenomenon of induced decomposition that occurs at concentrations above 0,1 M (2).

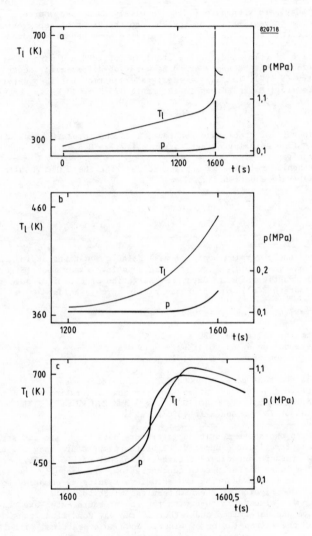

Figure 2.   Temperature and pressure course during
a thermal explosion experiment in the
low pressure autoclave at a prepressure
of 0,1 MPa.
a.   linear heating followed by thermal explosion
b.   thermal runaway stage of the thermal explosion
c.   explosion stage of the thermal explosion

4. Initiation stage of TBPB

The term thermal explosion is used in a manner that in slightly different from that in literature (8,9). The following scheme illustrates our definition:

THERMAL EXPLOSION	=	THERMAL RUNAWAY	+	INITIATION	+	EXPLOSION
time scale (seconds)		10^4-1		$10-0,1$		$1-10^{-3}$
main feature		uniform temperature		non-uniform temperature		propagating reaction zone through a running-away mass

In literature only the process leading up to an explosion is meant when the term thermal explosion is used and the explosion phenomena are not considered, since only critical temperature, critical mass and induction period are treated. However, this thermal explosion does not occur throughout the entire reaction system.

The initiation stage follows the thermal runaway stage, starts at a temperature difference of 2 K (arbitrarily chosen) between the thermo-couples and is characterized by a non-uniform temperature (up to strong temperature gradients) in the reaction system. It could be described as a reversed mixing process. Natural fluctuations of the temperature are amplified as a consequence of the coupling between temperature and heat generation of the exothermal reaction. As in mixing several diffusion modes can be considered: molecular, eddy and bulk diffusion. During the initiation stage liquid clumps develop which are higher in temperature than the average bulk temperature. Instead of liquid clumps also liquid--gas dispersion clumps - of which several regimes are shown in Figure 3 - may develop. The half-lifetime of these clumps in proportion to the half-lifetime of the reaction shows that concentration differences exist during this stage of the thermal explosion.

The initiation stage changes into the explosion stage as soon as a propagating deflagration zone comes into being. The dispersion regime of the reaction system may differ considerably at the moment of this change, as is illustrated by the photographs A and B. If TBPB is heated slowly from room temperature in a beaker at atmospheric pressure deflagration develops from the liquid-gas boundary layer while the bulk of the liquid shows only a few gas bubbles. The temperature is then about 370 K (photograph A). Fast heating of TBPB results in a deflagration that comes into existence when the temperature of the bulk is about 460 K and a large amount of gases is formed (photograph B).

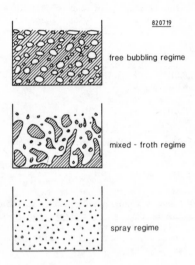

Figure 3. Several dispersion regimes.

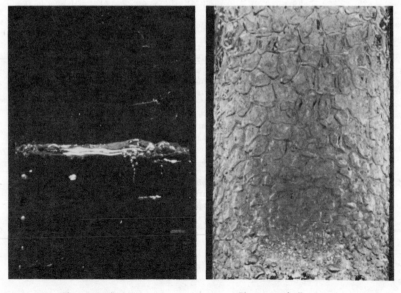

Photograph A   Photograph B

5. Explosion stage of TBPB

The intensity of the explosion is strongly related to the dispersion regime at the moment of initiation, viz. the moment when a deflagration starts. This becomes clear when considering Figure 4: the deflagration rate u suddenly increases when the reaction system changes from a free bubbling regime into a mixed-froth regime.

Figures 5 and 6 show the pressure and temperature curves of the thermal explosion of 100 g of TBPB in the low pressure autoclave at different prepressures. The heating rate was 44 mK s^{-1}, the gas temperature 300 K and the speed of the stirrer 5 r.p.s. At a prepressure of 0,8 MPa (nitrogen) the steep rise of the pressure coincides roughly with the moment when the deflagration zone is between the thermocouples. Deflagration takes place in the free bubbling regime. At the prepressure of 10 kPa the maximum pressure is almost reached when the temperatures in the sample vessel (T_{l1} and T_{l2}) rise steeply. Therefore the pressure course stems from the deflagration of a spray which is formed by thermal runaway.

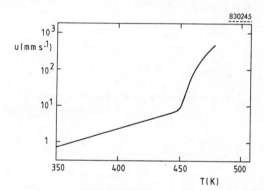

Figure 4. Deflagration rate u of TBPB at atmospheric pressure as a function of temperature T.

The large difference in intensity of the thermal explosion follows from a comparison of the values for the maximum pressure p_m and the maximum rate of pressure rise $\dot{p}_m$.

	$p_m - _o$ (MPa)	$\dot{p}_m$ (MPa s^{-1})
P_o = 0,8 MPa	2,05	75
P_o = 10 kPa	0,89	17

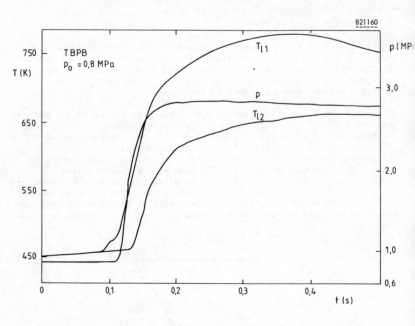

Figure 5.   Thermal explosion of TBPB at a pressure of 0,8 MPa.

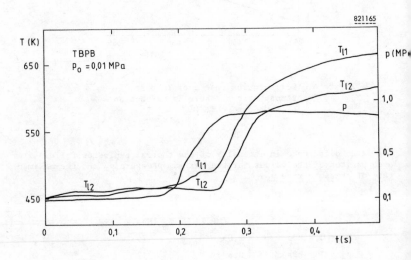

Figure 6.   Thermal explosion of THPB at a prepressure of 10 kPa.

6. Discussion

The scheme presented in paragraph 4 is a qualitative description of the different phenomena during a thermal explosion of a liquid. It has been proven that the behaviour of many other chemical substances is similar to that of TBPB (1).

The intensity of explosion is largely determined by the deflagration rate, which is a function of temperature, pressure and dispersion regime. The particularity of a thermal explosion of liquids is the propagation of a reaction zone through a running-away mass. Just as in deflagrations of hydrocarbon-air mixtures gasdynamic effects play an important role in thermal explosions.

The quantity of substance involved in a thermal explosion is very important. Extrapolation of test results from laboratory scale to amounts of 100 kg or 1000 kg is still very difficult in terms of explosion effects, because the deflagration phenomena during a thermal explosion have only been investigated to a limited extent.

The transition of deflagration to detonation during a thermal explosion is recommended as an object of study in the near future. A systematic investigation that takes account of the heat of explosion and the deflagration rate as a function of pressure, temperature and density (dispersion regime) seems a useful step in revealing some features of the very complex phenomena during the explosion stage of a thermal explosion of liquids.

References

1   J.Verhoeff   Experimental study of the thermal explosion of liquids, Thesis, Delft University of Technology, 1983

2   D.Swern   Organic peroxides, Vol I (1970), Vol II (1971), Vol III (1972) Wiley Interscience, New York

3   A.T.Blomquist, I.A.Bernstein   J.Am.Chem.Soc., 73, 1951, 5546

4   W.A.Pryor, E.H.Morkved, H.T.Bickley   J.Org.Chem., 37 (12), 1972, 1999

5   C.Filliatre, B.Maillard, J.J.Villenave   Thermochim. Acta, 39, 1980, 195

6   B.Maillard, J.J.Villenave, C.Filliatre   Thermochim. Acta, 39, 1980, 205

7   J.J.Villenave, C.Filliatre, B.Maillard   Thermochim. Acta, 39, 1980, 215

8   P.Gray, P.R.Lee   Oxidation and combustion reviews Vol 2, Ed.C.F.H.Tipper, Elseviers Publ. Comp., Amsterdam, 1967

9   A.G.Merzhanov, V.G.Abramov   Propellants Expl., 6, 1981, 130

EXPERIMENTAL PRINCIPLES FOR PREDICTING SAFE CONDITIONS FOR THE STORAGE OF BULK CHEMICALS

G. Giger, R. Gygax, F. Hoch

CIBA-GEIGY Ltd., Basle, Switzerland

INTRODUCTION

Risk analysis in chemical plants requires the gathering of information on the desired and undesired chemical reactivity of the materials to be processed. Because chemical reactions are intrinsically accompanied by thermal effects and because the risk to be assessed mainly stems from the thermal potential of the chemical transformations, data collection on decomposition reactions is mostly done by thermal measurements. Since these measurements are preferably performed on a small laboratory scale, the task of interpreting their results is mainly a problem of scale-up.

For the consideration of this scale-up two cases can be discerned:

(1) Synthetic reactions and subsequent conditioning operations in the chemical plant.

Two distinct advantages for controlling heat evolutions govern these situations: -1- The plant equipment is normally designed to allow for the dissipation of quite a quantity of heat and -2- there is always the possibility to monitor the temperature in order to get some on-line information on the thermal nature of the overall system, which makes corrective action possible (see also Ref. 1).

(2) Bulk storage and transportation of chemicals.

Assessment of the risk due to unintended reactivity is more difficult in the context of storage and transportation in two respects: -1- The heat dissipation in unstirred bulk materials may be exceedingly small. Therefore, due to a large capacity of the bulk for heat accumulation, heat evolution rates so small as to be hardly detectable in the laboratory experiment may still be important (2). -2- During storage and transportation most often no observation nor corrective measures are possible. The goods are removed from the control of the persons responsible for and knowledgeable of the safety aspects, i.e. absolute proof of the stability of the transported system must be given ahead of time.

SCALE-UP PREDICTIONS OF CRITICAL CONDITIONS

In this paper we would like to address the problem of performing scale-up predictions for the second case. The question of interest concerns the course of the temperature as a function of time for a considered system. Ultimately one wants to determine a critical temperature above which a thermal runaway of a given compound in a given container is predicted to just occur.

Criticality of reacting systems has been dealt with (3-6) often with a theoretical point of interest. We are concerned here with obtaining laboratory data and interpreting them to obtain predictions for larger scales.

In the case of unstirred systems, scale-up predictions are not trivial. The reason is that the local temperature distribution within the bulk and its course as a function of time is a result of the balance of two factors, which are affected very differently by scale-up:

a) The heat dissipation rate from the bulk to the surroundings strongly (i.e. quadratically) decreases with the increasing distance between the origin of the heat and the surface of the body.

b) The heat production rate is an intensive property of the involved compound and does not depend on shape and size of the material. It is, however, influenced by the actual temperature distribution within the bulk and, due to chemical conversion, by the previous thermal history.

The critical condition can be defined as the case where the two factors exactly balance. It is stationary only if chemical conversion is negligible during the response time of the heat conduction in a body of given size. If, in addition, the heat transfer through the surface wall is fast compared to heat transfer within the bulk, the following formulae, derived from Frank-Kamenetzkii's theory (3) and the Arrhenius law, hold:

$$\dot{Q}_{critical} = \delta_c \frac{\lambda}{\rho r_o^2} \frac{RT_{critical}^2}{E_a} \; ; \quad \dot{Q} = \dot{Q}_\infty e^{-E_a/RT} \qquad \text{I}$$

$\dot{Q}_\infty$ and E_a determine the heat production rate ($\dot{Q}$) and its dependence on temperature. The critical heat dissipation rate ($\dot{Q}_{critical}$) is related to the critical ambient temperature $T_{critical}$, by E_a, by the thermal conductivity λ, by the density ρ, by the constant δ_c (4,6) related to the shape of the body and by the inverse of the squared characteristical dimension r_o.

If all these quantities were separately known, the critical conditions could immediately be calculated for bulks of any size.

Several experimental approaches can be used to obtain enough data to
predict critical temperatures for a given bulk compound. Beever and
Thorne have shown an elegant way to such scale-up predictions based
on experiments determining the critical condition of several laboratory size set-ups (5). Other methods rely on experiments designed to
determine the heat production rate separately.

EXAMPLE CASE

In the following pages we should like to discuss critically these
methods always using the same compound as an example. As the larger
scale situation we choose the storage of the considered compound in a
100 l drum (height: diameter = 1.5:1).

Differential thermal analysis (DTA)

To begin with we obtain an overview of the thermal potential of the
compound by inspecting the scanning differential analysis thermogram
of fig. la. It consists of several exothermal signals with a total
energy release of about -1200 kJ/kg. An isothermal DTA-run at 80 °C
(fig. lb) releases about half of the total energy within 60 hours.
From these data we take the information that the decomposition reaction controlling the energy release rate in the low temperature region
involves an overall reaction energy of around -600 kJ/kg.

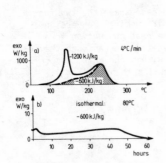

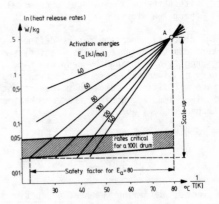

Fig. 1: Scanning (a) and isothermal (b) DTA of the
considered compound
(shaded area = thermogram
after experiment b)

Fig. 2: Relation of heat release
rates accessible experimentally (for example point A)
to range critical for a
100 l drum

Qualitative scale-up by an empirical safety factor

The dependence of heat production rates on temperature is most often
represented by a plot of their logarithm as a function of the inverse
absolute temperature (fig. 2). The initial heat production rates of a
single reaction step fall on a straight line. Deviations from the line
can be caused by reducing the reaction rates as a function of reagent
depletion or by several competitive reaction channels being active
under the experimental circumstances.

A laboratory experiment with a given sensitivity will only have access
to the upper part of a diagram like fig. 2. Correspondingly, there
exists an on-set temperature, above which the experiment starts to
detect the reaction investigated. For instance, the isothermal DTA
experiment of fig. 1b run at 80 ^{o}C reflects about the limit of the
sensitivity of such an experiment. 5 W/kg is an upper estimate for
the heat release rate at 80 ^{o}C; this nails down the single point $\underline{A}$
in fig. 2.

On the other hand, the range of heat production rates critical for
the drum can be estimated using typical values for λ, ρ and E_a in the
formula above. The range given in fig. 2 covers a 50 % correction of
the estimated values. The gap between point $\underline{A}$, which is accessible
with a laboratory experiment, and the range of rates relevant on the
larger scale must be bridged by the knowledge or assumption of the
temperature dependence of the heat production rates, i.e. of the activation energy E_a. This is done for several values in fig. 2.

In many cases it is sufficient to obtain a conservative estimate by
assuming a low value of E_a. Thus every degree of scale-up translates
into a safety factor which is deduced from the experimentally observed
on-set temperature.

In reality, safety factors have often been determined by experience
rather than explicitly in terms of the above analysis, which clearly
suggests, however, that safety factors cannot be generalized. They
not only depend on the scale to be judged upon but also on the nature
of the experiment on which the prediction is based. Obviously, the
degree of scale-up which can be covered by an overall safety factor
is limited, unless the safety factors become prohibitively large.

Many cases can be comprehensively dealt with using a safety factor,
if the actual temperature is sufficiently below a temperature determined to be critical under conservative assumptions. In practice, this
reduces the number of cases which have to be treated by more elaborate
means.

In the case of our selected example (fig. 2) the qualitative argument
using a reasonable safety-factor would require storage temperatures
well below 0 ^{o}C, a conclusion which is much too conservative. More
elaborate experiments are required to narrow in the critical temperature range.

Determination of critical conditions on the laboratory scale

If point $\underline{A}$ in fig. 2 can be determined such that it corresponds to the critical condition also on the laboratory scale, scale-up can be performed without knowledge of any constants in formula I except the activation energy E_a, which can be determined from criticality experiments by the method of ref. 5. This was attempted for our example by narrowing in the critical ambient temperature for baskets shaped as a regular cylinder of various sizes. While the baskets were kept in a fan-assisted oven, the temperature in the center of the basket was monitored as a function of time. Selected resulting curves for the 200 ml basket are shown in fig. 3.

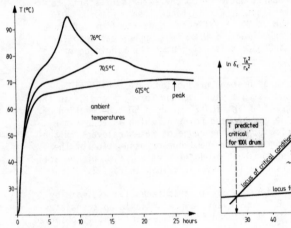

Fig. 3: Isoperibolic experiments in 200 ml basket

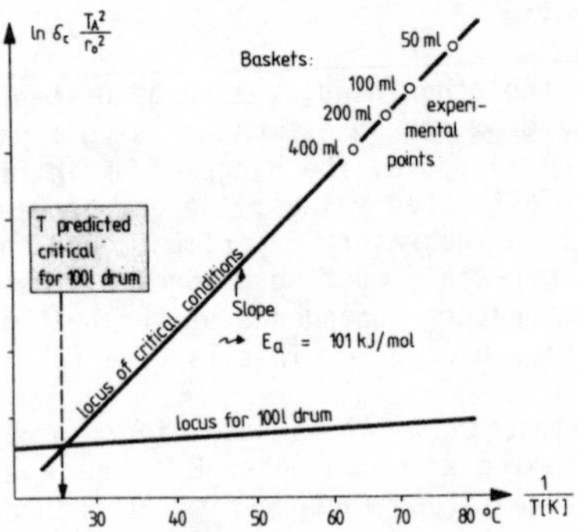

Fig. 4: Scale-up of critical conditions according to ref. 5

While for a true zero-order reaction an obvious switching of the shape of the curve is expected, in our case it can be clearly seen that eventually reagent depletion becomes dominant and the assumption of zero-order kinetics is, at most, only an approximation to the initial part of the curves.

Therefore, the definition of a critical temperature is not obvious. In line with curves obtained theoretically (6) we pick the curve which exhibits an accelerating phase, i.e. which does have a positive curvature in the selfheating phase, to be the first supercritical one. Thus we obtain the critical temperatures 78, 72, 68.5, 63.5 $^{\circ}$C (within $\pm 1 ^{\circ}$C) for 50, 100, 200 and 400 ml baskets, respectively.

According to ref. 5 formula I can be combined and linearized in the way shown in fig. 4. In this figure the above data points are tentatively plotted and a critical temperature around 30 ^{o}C is predicted for the 100 l drum.

Two points arise which put the legitimacy of this procedure in question:

- Since the procedure of ref. 5 assumes zero-order kinetics, what influence does a deviation from zero-order have on the critical temperatures? Since small material sizes behave less adiabatically, more heat evolved is lost to the surrounding as compared to larger bulks, i.e. a large degree of conversion will take place, before the temperature can rise into a critical region. Therefore, critical temperatures will be too high in smaller samples and the critical temperatures extrapolated to larger scales will be too high. We have performed some numerical calculations using parameters describing more or less the kinetics of the reaction (ΔH = -600 kJ/kg) and heat conduction in the experimental baskets. They show that in our case the critical temperatures only marginally depend on the assumed reaction order of either one or zero, and we conclude, that this system can nearly be described by zero-order assumptions. In general, however, the deviation from non-zero order kinetics will influence the predictions in the dangerous direction.

- The second deviation from the assumed conditions, however, points in the other direction. Since the Frank-Kamenetzkii regime, i.e. the predominant control of the heat dissipation by conduction through the bulk as compared to transfer at the surface, is more likely to be violated in smaller scales, their critical temperatures tend to become too low as compared to the ideal ones. Our numerically calculated estimates suggest that this influence may be realistic in our experiments whose scale-up predictions of critical temperatures are therefore considered as being on the lower side.

EXPERIMENTS DETERMINING THE HEAT
PRODUCTION TERM SEPARATELY

Obviously it would be advantageous if the thermo-kinetic properties could be determined separately from any influence of heat dissipation. In principle, this can be achieved in two ways:

- In an <u>adiabatic experiment</u> all the heat produced is retained and no heat is dissipated. The temperature increase curve directly bears the kinetic information required.

- In an <u>isothermal experiment</u> all the heat produced is immediately dissipated, i.e. the measured heat flux from the sample container to the surrounding is equal to the heat produced.

Unfortunately both adiabatic and isothermal conditions are idealisa-
tions which are fairly hard to realize for an unstirred bulk over a
long enough time.

Adiabatic thermal measurement

In some instruments the adiabatic state is approached by equalizing
the temperature of the surrounding oven with the sample temperature
by electronic control and therefore avoiding heat flux between the
oven and the sample. Measurements of typical samples including volatile components require pressure tight containers whose relatively
heavy mass acts as a heat sink. During the self-heating process this
heat sink disturbs the temperature profile within the sample, causing
heat to flow and violating the adiabatic requirement. The Accelerating
Rate Calorimeter (ARC, ref. 7) and the concept of interpretation of
its results takes into account these difficulties to some degree.

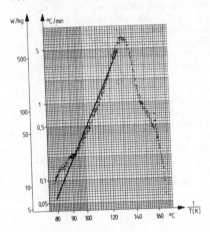

Fig. 5 displays the first peak
of an ARC-run of the example
compound. The reaction typically does not show features
clearly corresponding to a
single kinetic model. However,
neglecting the initial points,
an estimate of the temperature
dependence of the main reaction is possible. The line
drawn in fig. 5, which presumably gives a lower estimate
for the "zero-order slope" (7)
corresponds to an activation
energy E_a of 105 kJ/mol.

Fig. 5: ARC run of the considered
compound (= 3,85, ref. 7)

Isothermal measurements

In an unstirred system isothermal conditons can in principle not be
maintained within a heat producing sample. In order to remove the heat
a temperature gradient must be established across the sample. Isothermal conditions can be approximated if the time scale of the heat
removal is small compared to the time-scale of interest and if the
temperature deviation between the center of the sample and the surroundings is negligible. This can be achieved using very small samples
(Microthermoanalysis) or measuring in the low range of heat production
rates. In terms of the achievable sensitivity these two conditions are
mutually exclusive. In the following an indirect method is advocated
which avoids this dilemma.

Combined thermal and chemical analysis

From fig. 1 we did conclude, that the rate determining decomposition reaction was related to an energy conversion on the order of -600 kJ/kg.

In separate experiments, series of samples were exposed to different temperatures in containers shaped to allow for good heat exchange with the heat bath. At various intervals the assay of the compound was determined by chemical analysis. Samples of the thermally treated materials were also run on scanning DTA equipment indicating the energy content of the first signal to decrease parallel to the decrease of assay. Results of the analytical measurements are shown in fig. 6. The initial decay rate is approximated by a straight line through the data points.

By utilizing the overall heat of reaction, the assay decay rates can easily be transformed into the corresponding heat production rates and vice versa. Similar to the argument used in fig. 2 a range of critical assay decay rates can be fixed for the 100 l drum. This range is plotted in fig. 7 together with the experimental data points in a fashion familiar from the previous discussion. From the slope of the straight line an activation energy of 108 kJ/mole is obtained and the critical temperature is predicted to be around 40 $^{\circ}$C.

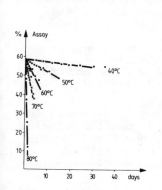

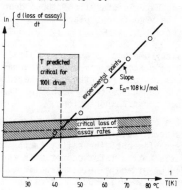

Fig. 6: Assay decay rates by thermal treatment

Fig. 7: Temperature dependence of initial assay decay rates and rates critical for a 100 l drum

Although this procedure is based on certain assumptions also, it has several advantages:

- Considering the initial decay rates, it is assured that the zero-order assumption is valid in the range of the experimental points. If reagent depletion affects the prediction, it can only do so in the conservative direction.

- In this treatment, the temperature of interest can actually be covered. By extending the thermal treatment to lower temperatures, every desirable "sensitivity" of such an indirect method can be achieved. This avoids the necessity of an extrapolation to lower temperatures thereby eliminating an important factor of uncertainty.

- Although the reaction mechanism can, to some degree, still be treated in quite a global manner, some insight into its influence on the thermal behaviour of the compound is necessary. Especially if the reaction exhibits self-acceleration even under isothermal conditions (ref. 8), the induction times must also be taken into account. The gathered experimental data do provide evidence of such a mechanism, if and when it becomes important.

CONCLUSIONS

Decomposition reactions of industrial samples do not normally allow for a clean quantitative description unless a considerable effort is extended. We have shown, at least with one example, that by critical inspection of the results of various simple experiments, whose predictions of E_a and the critical temperatures were remarkably close, an adequate knowledge of the thermal properties of a compound can be obtained. Experiments with a 100 l drum are under way and, consistently with the prediction, place the critical temperature in the neighbourhood of 40 °C.

By using numerical simulation calculations or simple arguments based on these data, the compound's thermal behaviour under various circumstances can be predicted. We like to conclude with some suggestions for the design of tests assessing the hazard of self-heating during storage and transportation:

a) If a compound exhibits exothermal behaviour only well above temperatures realistic for the actual storage, a simple test procedure is sufficient. The safety factor must be chosen on an educated basis taking into account both the quality of the test and the degree of scale-up. In practice, many cases can be dealt with comprehensively in such a screening manner.

b) If the safety of storage or transport cannot be sufficiently guaranteed by this procedure, laboratory experiments should be performed which both explore the heat production rate at a given temperature and its temperature dependence. This way of proceeding is better than to perform a single, even one-to-one scale experiment, which determines just one point of the diagram used above. While the latter procedure can be used for verification of scale-up predictions it does not, by itself, provide enough information on the probability of a runaway under real conditons. The former procedure, on the other hand, makes possible an estimate on how large a deviation, say by catalytic effects, by an unintended exposure to heat or the use of larger packing units must become, before a stable condition is turned into an unstable one.

REFERENCES

1. P. Finck, G. Giger, H. Fierz, R. Gygax, this symposium

2. F. Brogli, 3rd Int. Symp. Loss Prevention and Safety Promotion in the Process Industries 5/369 (1980)

3. Reviewed in: P. Gray & P. R. Lee, Oxid. Combust. Rev. $\underline{2}$, 1 (1967)

4. P. C. Bowes, The Inst. Chem. Eng. Symp. Ser. 68, 1/A:1, Rugby, England (1981)

5. P. F. Beever & P. F. Thorne, loc cit 1/B:1

6. T. Boddington, P. Gray & I. K. Walker, loc cit 1/C:1

7. D. I. Townsend & J. C. Tou, Thermochimica Acta 37, 1 (1980); D. I. Townsend, The Inst. Chem. Eng. Symp. Ser. 68 3/Q:1, Rugby, England (1981)

8. F. Brogli, P. Grimm, M. Meyer, H. Zubler; 3rd Int. Symp. Loss Prevention and Safety Promotion in the Process Industries; 8/665 (1980)

Pressure increase in exothermic decomposition reactions,
Part II

O. Klais and Th. Grewer

Hoechst AG, 6230 Frankfurt 80, Germany

SUMMARY
The risk associated with exothermic decomposition reactions is
the pressure build-up in a closed vessel. An improved measurement facility was employed and it was shown that complete decomposition of many organic compounds proceeds via several reaction stages to the thermodynamically preferred stable products.
An exponential rise in pressure as predicted by the theory of
thermal explosions only applies for the initial phase of decomposition. Upon more intense self-heating the decomposition
rate approaches a limiting value. For substances with the tendency toward deflagration, the pressure rise is different and
follows a power function.

1. Introduction
The purpose of recording pressure-time curves of decomposition
reactions is to contribute toward systematic understanding of
the hazards arising from such unwanted reactions. In these
studies our aim was to achieve laboratory simulation of the
situation in which decomposition is no longer under control
and the hot gases released by the reaction lead to a pressure
build-up in a closed container. We reported on the first results at the last symposium (1).

2. Experimental work described in the literature
Although the pressure build-up during the deflagration of an
explosive is a characteristic property which is frequently
reported in the literature, we know of only two reports of
experimental work in which the explosion pressure and the
associated pressure build-up rate of a homogeneous thermal
explosion were determined. For the classification of peroxides
Groothuizen et al. (2) use a 10-l pressure-proof vessel within
which a 100-ml test vessel is heated up under temperature controlled conditions. The pressure curve in the pressure vessel
is recorded. Berthold (3) describes a test set-up in which the
pressure vessel can be filled to a higher degree with the test
substance. However, the pressure vessel is not designed to
withstand the pressure of a decomposition reaction which runs
to completion, so that it must be equipped with a bursting
disk for pressure venting.

Test methods which were developed to obtain caloric measurements and later adapted also to record pressure we consider
unsuitable for determining the pressure build-up of a thermal
explosion. The heat flow to the vessel walls under conditions
of large surface area/volume ratios impedes the self-accela-
ration of the decomposition process and the rate of pressure
build-up measured is reduced accordingly.

3. Experimental part

The experimental set-up used is described in (1), but was
substantially improved for the trials reported here: the
pressure vessel has the form of a cylinder with the height
equal to the diameter; the test pressure was made high enough
to permit a higher degree of filling with the test substance
(g substance/cm^3); placing the pressure vessel in a drying
oven with forced air circulation permits homogeneous temperature control of the pressure vessel along with the valve and
pressure transducer. Fig. 1 shows the design scheme. From the
scheme it can be seen that it is possible to work with two
pressure transducers. Since the piezoelectric measuring system
can record only pressure changes and does not permit static measurement of the pressure curve, for example during the cooling
phase, an added measuring system for static pressures was provided to determine the amount of gas formed.

Table 1: Pressure vessels used (Nova company)

Volume		Test press.	Intern. diameter		Height	Weight
without glass insert cm^3	with glass insert cm^3	bar	pressure vessel mm	glass insert mm	mm	kg
200	180	700	55	50	85	8.8
100	80	2000	35	30	120	11.6

To avoid catalytic reactions of the test substance with the
vessel walls and for better thermal insulation the samples
were alway first placed into a glass vessel that closely fitted
into the pressure vessel.

The entire test set-up was heated up under programmed control.
The temperature curves for the oven, the pressure vessel(thermocouple in lid) and the sample are shown in Fig. 2 for the
example of the decomposition of 2-nitrobenzaldehyde. The melting of the sample is easily recognized. During the slow hea-
ting-up period (0.6 K/minute) following melting, equilibrium
is established between sample temperature and pressure vessel
temperature, which enables one to estimate the heat flux. In
the example a heat equivalent corresponding to a temperature
change of 1 K per minute per 1 K temperature difference flowed
into the sample. Similar values were determined for other substances above their melting points. In contrast the heating up
rate measured in the centre of powder samples amounted to only
0.1 K per minute per 1 K temperature difference.

On the other hand, when the sample is heated above the pressure
vessel temperature, a similarly large quantity of heat flows
into the pressure vessel. This results in a damping or, in the
extreme case, suppression of the thermal explosion. This should
be particularly pronounced when the heating rate is low.

Recording techniques were also substantially improved in comparison with the set-up in (1) through the use of a transient
recorder with higher resolving power (Bryans, type 523 A) and

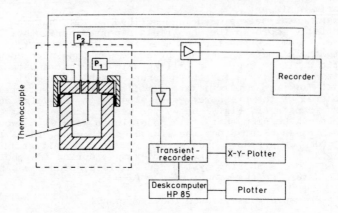

Fig. 1 Schematic diagram of the experimental set up for the registration of the pressure increase in exothermic decomposition reactions.

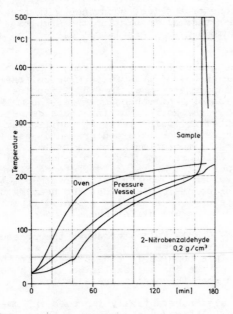

Fig. 2 Temperature curves for the oven, the pressure vessel and the sample

a desk-top computer (HP 85 with plotter).

4. Results

Our primary interest in making the measurements was the search for recognizable relationships between experimental parameters, such as degree of filling, heating rate and pressure vessel shape, and the decomposition characteristics. We were limited in our choice of test substances because, owing to the design of the measuring system, the dynamic pressure transducers (natural frequency $\geq$ 200 kHz) did not permit temperatures $>$ 240°C for the high pressure range ($>$ 300 bar).

For the first time it was possible to detect in the recorded pressure-time curves a dependence of the pressure increase rate on the rising pressure, which indicated a multistage decomposition process. As an example Fig. 3 shows the pressure-time curve for the decomposition of 3-nitrobenzaldehyde. In the curve three contrasting decomposition rates can clearly be seen.

The differing character of the decomposition reactions is more clearly revealed by a plot of the pressure increase rate versus the pressure in the vessel than by the directly measured pressure-time curve, as is shown for the three isomers of nitrobenzaldehyde in Fig. 5. With qualifications such a diagram is comparable with the thermogram of a differential thermoanalysis, if pressure is taken as a measure of heating and the pressure increase rate as a measure of heat production. Such a multistage course of decomposition was not observed only in the example shown but occurred in most of the samples so far examined. Almost without exception the highest measured pressure rate is found in the first reaction stage.

The results of our studies to characterize a decomposition reaction are summarized in Table 2. The maximum pressure and the pressure increase rate are also shown in relation to the degree of filling in order to yield better comparability and applicability to other containers.

5. Discussion

From trials on the thermal stability of the chosen substances it is known that in its initial stage their thermal decomposition can be described by a first order reaction. It is also an obvious assumption that the amount of gas generated by decomposition is proportional to the heat energy developed. For a thermal explosion that proceeds in this way one anticipates an exponential pressure increase that flattens out at higher pressures as substance is consumed.

Curves with the expected form were found for azoisobutyronitrile, Fig. 4 , and the other azo compounds. More careful analysis showed that during decomposition the amount of gas generated exactly corresponds to the amount of nitrogen in the molecule. For some samples nitrogen could also be detected as decomposition gas.

To provide additional evidence we use a simple heat balance for the decomposition of azoisobutyronitrile. Of the decomposition energy measured of 213 kJ/mol the gaseous products generated absorb ca. 35 kJ/mol to build up the recorded pressure of 130 bar. Under adiabatic conditions the remaining

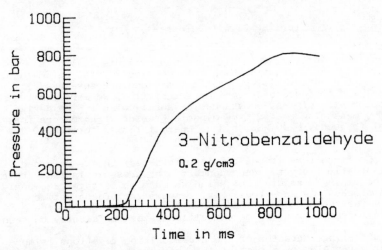

Fig. 3 Decomposition of 3-nitrobenzaldehyde: recorded pressure-time-curve

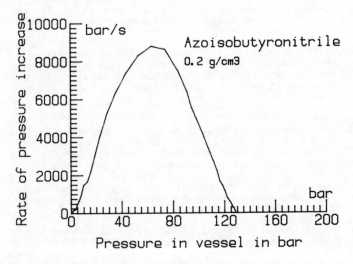

Fig. 4 Decomposition of azoisobutyronitrile: pressure increase rate versus pressure in the vessel

test substance with a specific heat of 2 J/g.K should heat up
by about 660 K. Because of the unavoidable heat losses this
temperature increase is hardly achieved in the experiment. The
increase of 310 K to 390 °C recorded with a thermocouple probably comes closer to reality.

Whether a decomposition reaction runs to completion to form
thermally stable products (e.g. H_2O, CO, N_2), thus depends
essentially on the energy released in the first decomposition
stage and the temperature difference between the first stage
and successive decomposition reactions. If a product of considerable greater thermal stability is formed, decomposition will
be interrupted before the reaction can proceed to completion.

DTA thermograms recorded with heating rates comparable with
those chosen here exhibit for the three isomers of <u>nitrobenzal-
dehyde</u> several partially overlapping exothermal reaction peaks
in the temperature range from 190°C to 350°C. For the meta-
and para-configurations three separate stages are clearly
discernible, but for the ortho-configuration the decomposition
reactions cannot be so readily distinguished from one another
(Fig. 6). On the basis of the information supplied by thermoanalysis a multistage process is expected for the decomposition of nitrobenzaldehyde and is found in the curves shown in
Fig. 5. The higher pressure build-up also indicates complete
decomposition of the samples to volatile products, and in fact,
when the pressure vessel is opened, the only residue found is
a soot film on the vessel wall.

Decomposition of the nitrobenzaldehydes is presumable initiated
for all three isomers by the aldehyde group. To show this
clearly, in Fig. 7 the logarithm of the pressure increase rate
is plotted versus pressure for the three isomers. According to
the theory of thermal explosions such a plot should yield a
linear function if it is assumed that the pressure increase
is proportional to the temperature increase:

$$\frac{dp}{dt} \propto \frac{dT}{dt} = A \cdot \exp\left[E/RT_0^2 \cdot (T-T_0)\right] \propto \exp\left[a \cdot (p-p_0)\right]$$

> a= adaptation factor influenced by the moles of gas
> generated per mole of solid, in addition to the
> activation energy.

The slowing down of the pressure increase rate with increasing
amount of material reacted was already emphasized by us in (1).
The exponential increase of the reaction rate assumed in the
theory is apparently limited to cases of small temperature increase.

A reduction of the reaction rate with increasing temperature
was also observed for the decomposition of explosive substances (4). A plausible explanation for this observation is that
an endothermal dissociation reaction with high activation
energy initiates decomposition and draws its energy from exothermal secondary reactions. As the amount of reacted substance increases, the rates of the secondary reactions come to
determine the overall reaction rate. If their rates are lower,
the pressure increase slows accordingly. How much the course of

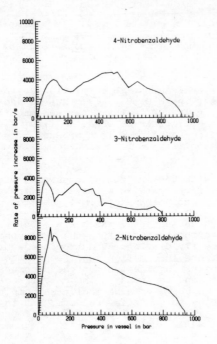

Fig. 5 Pressure increase rate versus pressure in the vessel for the three isomeres of nitrobenzaldehyde.

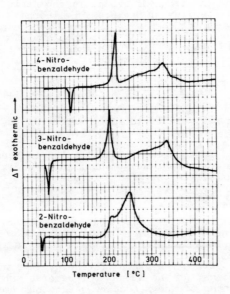

Fig. 6 DTA thermograms for the three isomeres of nitrobenzaldehyde

the reaction is determined by chemical structure is shown by
the difference between the measurements of 2-nitrobenzaldehyde
and those of the other isomers. The overlapping of the individual
reaction stages revealed by differential thermoanalysis for
2-nitrobenzaldehyde makes it seem likely that more heat is released in the first stage of decomposition. Fig. 5 shows similarly a higher pressure increase rate.

If a particular product distribution is assumed, the heat released by decomposition can be calculated from the thermodynamic data. If it is assumed that 3-nitrobenzaldehyde decomposes to $5/2$ H_2O + $1/2$ CO + $1/2$ N_2 + 6 $1/2$ C, the maximum heat
of reaction amounts to 540 kJ/mol. In the experiment it was
found, however, that 2-3 moles of uncondensable products were
found for each mole of solid, which contradicts the assumed
product distribution. The decomposition energy measured by DSC
amounted to 220 kJ/mol.
In contrast with the other substances ammonium dichromate tends
to deflagrate at room temperature, that is to say, if a pile of
the powder is ignited a decomposition zone propagates through
the sample with intense formation of nitrogen. Trials in our
laboratory have shown that the burning rate rises as a power
function with increasing pressure and the rate of gas development is best represented by the expression:

deflagration of ammonium dichromate: $\frac{dp}{dt} \propto p^{1.6}$

The proportionality constant of the relationship depends only
on the degree of filling of the vessel. A log-log plot of the
pressure increase rate versus the pressure should therefore
yield a straight line.

A striking feature of the empirical pressure-time curves of
ammonium dichromate decomposition was that the pressure increase rate rose sharply from a value < 0.1 bar/s to values
> 10 bar/s and further accelerated with increasing pressure.
This behaviour was not observed for any of the other substances studied. The obvious explanation that a deflagration is
initiated by local overheating is confirmed by the log-log plot
of the pressure increase rate versus pressure in Fig. 8. The
curve found is that which is expected for a deflagration; the
slope of the line obtained when ignition is induced at room
temperature by an ignition source is identical with that obtained here where deflagration is initiated by thermodecomposition.

Under the assumption of quasi-adiabatic conditions the enthalpy
of decomposition is consumed by the internal energy of the gas
formed, by the external work performed and by the heating of
the solid residue.

$$\Delta H_{decomposition} = \Delta U_{gas} + \Delta U_{solid} + \Delta n.RT$$

In the decomposition reaction

$$(NH_4)_2 Cr_2O_7 \longrightarrow N_2 + Cr_2O_3 + 4H_2O$$

a heat of reaction of 174 kJ/mol is released and absorbed by
the gases (specific heat of H_2O = 42 J/mol.K; N_2 = 33 J/mol.K

for the temperature difference 500 - 1700 K) and chromium
oxide (specific heat 119 J/mol.K). The calculation employing
these data yields a temperature increase by 1440 K to a final
temperature of 1900 K. On the other hand the maximum temperature estimated from the maximum pressure measured and the
above products of decomposition in accordance with the ideal
gas law is 1610 K (test with degree of filling = 0.2 g/cm^3 and
510 bar maximum pressure), which is in reasonably good agreement with the maximum value calculated for adiabatic conditions.
It should also be mentioned that after the pressure vessel had
cooled, the pressure dropped as expected to the partial pressure of the N_2 formed.

6. Conclusion
An improved measurement facility was employed to investigate
the pressure build-up of decomposition reactions. Complete
decomposition of organic compounds proceeds via several reaction stages to the thermodynamically preferred stable products.
If the amount of energy released in the first stage of decomposition is small, decomposition comes to a halt before the reaction can proceed to completion.

A more careful analysis of the pressure build-up during the
initial phase of decomposition showed that an exponential rise
in pressure as predicted by the theory of thermal explosions
only applies where the pressure increase is very small. Upon
more intense selfheating of the sample the decomposition reaction rate approaches a limiting value.

For substances with a tendency toward deflagration, this process can also be initiated by thermodecomposition. The pressure
rise then follows a power function.

7. Acknowledgment
This paper has been partly supported by the German Ministry of
Research and Technology. The authors thank H. Smeykal for his
assistance in the experimental work.

8. References
(1) Grewer, Th. and O. Klais
    "Pressure Increase in Exothermic Decomposition Reactions"
    3rd International Symposium on Loss Prevention
    page 8/657 (1980)
(2) Groothuizen, Th.M.; J. Verhoeff and J.J. Groot
    "Determination of the Effect of a thermal Explosion of
    Organic Peroxides"
    J.Harzard.Materials 2, 11 (1977)
(3) Berthold, W.; K. Imbery and U. Löffler in
    "Die Berufsgenossenschaft"/March 1979
(4) Janswoude, J.J. and H.J. Pasman
    "Decreasing Progression of the Decomposition Rate of
    Explosive Substances at high Temperatures"
    in "Fast Reactions in Energetic Systems"
    p. 515 (1981)
    ed. C. Capellos and R.F. Walker

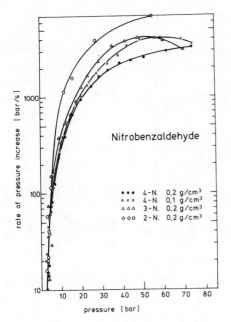

Fig. 7 Semilogarithmic plot of the pressure increase rate versus pressure for the three isomeres of nitrobenzaldehyde. An experimential increase is only observed at low pressure.

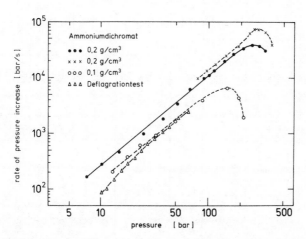

Fig. 8 log-log plot of the pressure increase rate versus pressure for the decomposition/deflagration of ammonium dichromate

Substance	Vessel/Pressure	Mass/Volume g/cm³	Pmax bar	Ps kPa.m³/kg	dp/dt max bar/s	dPs/dt kPa.m³/kg.s	Initial-Temperature °C	Heating rate K/minute
Diazonium salt	200 cm³/700 bar " " "	0.1 0.2 0.33 0.38	75 200 460 370	75 100 140 100	- 66 170 245	- 33 52 65	112 112 105 111	0.77 0.91 0.96 0.96
Azoisobutyro-nitrile	200 cm³/700 bar " "	0.1 0.2 0.4	59 130 340	60 65 85	4,580 8,800 (16,000)	4,600 4,400 (4,000)	95 80 93	0.5 0.9 0.5
1,3-Diphenyl-triazene	200 cm³/700 bar "	0.1 0.2	43 95	43 48	105 420	100 210	140 140	0.4 0.4
2-Nitrobenz-aldehyde	200 cm³/700 bar " " "	0.1 0.2 0.2 0.2	390 945 200 (40)	390 450 100 (20)	5,900 8,700 3,630 -	5,900 4,400 1,800 -	202 200 200 188	0.7 0.6 0.35 0.13
3-Nitrobenz-aldehyde	200 cm³/700 bar " 110 cm³/2000bar	0.1 0.2 0.2	370 830 830	370 410 410	4,500 4,100 3,600	4,500 2,100 1,800	190 190 200	1.0 0.9 0.7
4-Nitrobenz-aldehyde	200 cm³/700 bar " 110 cm³/2000bar	0.1 0.2 0.2	390 960 840	390 480 420	4,100 4,700 12,000	4,100 2,400 6,000	195 200 210	0.9 0.7 0.8
Ammonium dichromate	200 cm³/700 bar " "	0.2 0.2 0.1	475 510 220	240 250 220	38,000 68,000 6,300	19,000 34,000 6,300	217 227 226	0.1 0.2 0.4

Table 2: Maximum pressure and rate of pressure increase of decomposition reactions

Fire and Explosion

Chairman

D.H. Napier
University of Toronto,
Canada

Fire engulfment trials with LPG tanks with a range of fire protection methods

A F Roberts, D P Cutler and K Billinge

Explosion and Flame Laboratory, HSE, Harpur Hill, Buxton, Derbyshire,

ABSTRACT

A series of fire engulfment trials has been carried out on 500 litre LPG tanks. The tanks, containing 200 litres of fluid, were installed above a pool of kerosene surrounded by a 1m high windshield. The flames from the burning kerosene totally engulfed the tank and gave reproducible heating under a range of atmospheric conditions. The tanks were instrumented with thermocouples and a pressure tapping.

Trials were conducted with uninsulated tanks containing water, uninsulated tanks containing propane and insulated tanks containing propane with a range of insulating materials. In each trial data was obtained on heat transfer rates to the total system and the tank contents, the boiling regime and tank wall temperatures.

The pressure relief valve first operated after 3 minutes for an uninsulated tank containing propane, and in the range 12-90 minutes for insulated tanks with propane. Often the propane boiled in the film regime giving high tank wall temperatures below the liquid level; for the insulation systems with the best performance, boiling was in the nucleate regime with lower wall temperatures.

The average heat flux from the fire to the tanks was $130kW/m^2$ and was fairly uniform around the tank. Comparisons are made of test severity with those of tests elsewhere.

Introduction

The need for some means of protecting pressurised storage vessels against an external fire has long been recognised. In the absence of such protection, an external fire will heat the vessel contents causing a rise in internal pressure and heat the vessel walls causing a loss in mechanical strength; these effects may lead to rapid catastrophic failure of the vessel and a sudden release of the contents. Such a release can lead to blast effects, and if the contents are toxic or flammable other serious effects can result.

Methods of protection include the use of pressure relief valves, water sprays and thermal insulation. Relief valves are designed to open somewhat above the maximum anticipated service pressure and to vent the contents at a rate sufficient to stop the pressure rising appreciably higher than the operating pressure; disadvantages of relief valves include the possibility of mechanical damage (particularly on mobile systems) and the fact that thermal weakening of the vessel walls can still lead to violent failure at pressures less than the stable internal pressure for a correctly sized valve.

Water spray systems are designed to give a uniform distribution of water (usually at about 0.2 gall/ft^2. min or 0.16kg/m^2.s) over the surface of a vessel in the event of an external fire. This is designed to maintain the walls at a moderately low temperature, thereby maintaining wall strength and reducing heat transfer to the contents. Disadvantages of water sprays include the size and cost of large systems and reliability and maintenance problems.

Thermal insulation systems are designed to keep wall temperatures and heat transfer rates low by introducing a low thermal conductivity barrier between the fire and the vessel. Cement based insulation systems have been available for many years; their disadvantages include their weight and difficulties in inspection of the vessel. Burying vessels is a form of thermal insulation protection; disadvantages here include possible corrosion by ground water and possible explosion hazards if flammable vapours leak undetected and accumulate below ground.

Thus, each of the available methods of protection has advantages and disadvantages. The choice of a method for a particular situation depends very much on the detailed circumstances.

Little guidance is available on design criteria for insulation systems. Also, new insulation systems have recently been developed that appear to offer significant advantages over cement based systems, through substantial weight reduction for a given insulation performance. It was therefore considered desirable to conduct a series of fire engulfment tests on LPG tanks protected by various insulation systems, with the following objectives:

1. to gain information on the protection afforded by modern insulation systems

2. to develop a view on reasonable test methods and performance specifications for such systems.

3. to establish a data base on the overall behaviour of insulated LPG tanks under fire engulfment conditions.

There are many factors requiring consideration for a viable insulation system apart from good insulation performance under fire exposure, including weight, cost, durability and ease of application. These additional factors have been reviewed (1) and will not be discussed here.

Fire engulfment trials

These trials were intended to expose systems to conditions reasonably representative of those likely to be experienced in a large accidental fire.

A 500 litre LPG container was chosen since it was large enough for a realistic test of an insulation system and was easily obtainable. Such containers are about 2m long and of 0.5m diameter, and the test facility was designed around these measurements.

A rectangular pool with a firebrick lining was constructed with dimensions 4m x 2.4m x 0.5m deep to contain kerosene for the experimental fires. Two piers were provided in the pool to support the LPG containers. The lowest point of the container was flush with the top of the pool walls. On the ground outside the pool a 1m high screen was erected at a distance of 1m from the pool walls, thus enclosing an area 6m x 4.4m.

Preliminary trials with this facility showed that a fully developed pool fire was established in 30-90 seconds from ignition (the variability being due to weather conditions) and a steady kerosene burning rate of about 0.64 1/s ensued. The facility could therefore be used for fires of over 100 minutes duration. Without the screen present, the test tank was not fully engulfed by flames under all wind conditions, but with the screen present flame of at least 1m thickness engulfed the tank.

The experimental procedure was to attach 8 thermocouples to the walls of the LPG container at the positions shown in Fig 1 and then have the test insulation applied by a contractor. Three thermocouples mounted on a support rod were inserted into the container (Fig 2) and a pressure tapping was taken through a second opening in the container wall to a pressure transducer. A pressure relief valve, set to operate at 1.7 MPa (250 psi), was fitted, the container was placed on the piers in the pool and 200 litres of commercial propane were transferred into it.

The quantity of kerosene necessary to give a fire of the required duration was then run into the pool and ignited. Temperatures and internal pressure were recorded through the test and each test was videotaped.

Calibration trials using water

In order to obtain data on heat transfer rates to the test tanks, two trials were performed to the above procedure using 200 litres of water instead of propane and with no pressure relief valve or pressure transducer fitted. These trials gave closely similar results, demonstrating the reproducibility of the system in terms of the temperatures recorded by the thermocouples (Figs 3 and 4). These show that the average upper wall temperature reaches $400^{\circ}C$ in under 90s and $800^{\circ}C$ in 240s from ignition. It starts to level out approaching $950^{\circ}C$ but drops to a value fluctuating between $850^{\circ}C$ and $910^{\circ}C$ at 8 minutes. The average lower wall temperature rises to $150^{\circ}C$ at 120s, slowly rises to $200^{\circ}C$ at 8 minutes where it remains reasonably steady thereafter. The bulk temperature of the water as recorded by the bottom internal thermocouple, B, achieves a constant value of about $120^{\circ}C$ in 8 minutes. The middle thermocouple, M, records $100^{\circ}C$ at 3 minutes when B is only recording $30^{\circ}C$, rises slightly and from 8-11 minutes records the same as B. Thereafter it rises sharply and records a similar temperature to the top thermocouple, T, in the range $300-350^{\circ}C$. T, in the vapour space, rises to $430^{\circ}C$ after 4 minutes, where it remains roughly steady until 8 minutes, then drops to $250^{\circ}C$ from 10-15 minutes before rising again.

This sequence of variations may be described as follows:

The upper tank surfaces heat up rapidly while the lower surfaces, being in contact with water, are regulated in their temperature variations by the variations in water temperature. The vapour space is heated by the upper surfaces and, being hotter than the water, transfers heat to the water surface. This surface becomes hotter than the bulk of the water and begins to boil at 4 minutes.

At 8 minutes the dominant process of water boiling at the wall begins, due to heat transfer through the walls. The flow of steam into the vapour space between 4 and 8 minutes is relatively slow; it checks the temperature rise of the vapour space by mixing in relatively cool steam at 100°C but does not affect the upper wall temperatures. At 8 minutes, general boiling produces a much greater flow of steam which reduces the temperature of both the vapour space and the upper walls. Thereafter, conditions are relatively steady. There was only a small orifice for the steam to flow through. This results in a pressurisation of the tank and elevation of the boiling point. Temperature records are consistent with nucleate boiling.

A heat balance on an element of the upper portion (i.e. the non wetted portion) of the tank wall, at temperature T_1, may be written as follows (assuming negligible convective heat transfer within the tank):

$$Q - \sigma \varepsilon T_i^4 - \sigma F T_i^4 - \rho c d \frac{dT_i}{dt} = 0 \quad - - - - - - - - - - - - - - (1)$$

where the first term is the heat received from the fire (which is assumed to be constant throughout the fire), the second term is heat radiated externally by the wall, the third is heat radiated internally by the wall to the liquid layer and the fourth is the heat remaining in the walls (all terms per unit time and unit surface area). The time is t, ε is the emissivity of the external surface and σ is the Stefan-Boltzmann constant. F is a number between 0 and 1 which incorporates the effects of internal wall surface emissivity, configuration factor of the liquid layer surface with respect to the non-wetted portion of the wall and reflectivity/absorptivity of the liquid layer.

The solution to equation (1) may be written as

$$Ln\left(\frac{T_A+T_i}{T_A-T_i}\right) - Ln\left(\frac{T_A+T_0}{T_A-T_0}\right) + 2\tan^{-1}\left(\frac{T_i}{T_A}\right) - 2\tan^{-1}\left(\frac{T_0}{T_A}\right) = \frac{4Qt}{\rho d c T_A} \quad - - - - - - - (2)$$

where T_0 is the value of T_1 at t = 0, d, ,c = thickness, density and specific heat of the wall and T_A is the asymptotic value of $T_1 = \left(\frac{Q}{\sigma(\varepsilon+F)}\right)^{\frac{1}{4}}$

This solution was used to obtain estimates of Q and F from the wall temperature data in Fig 3 using the values of d = 5mm, ρ = 7830 kg/m³ and c = 490 J/kg. °C. The calculated line shown in this figure gives an excellent representation of the experimental curve assuming the asymptotic temperature in the absence of convective heat transfer (which is valid up until 8 minutes after ignition) to be 950°C. This line was calculated with Q = 134 kW/m² and F = 0.055. This value of Q is comparable with other estimates of heat transfer to objects engulfed by large fires and corresponds to radiation from a unit emissivity source at 966°C. The estimated average temperature of the flames surrounding the tank was 1100°C. The value of F = 0.055 is consistent with a unit emissivity inner surface, a geometric factor for radiative heat transfer of 2/π and a fraction of incident radiation absorbed by the water = 0.09.

Thus, at equilibrium, the wall temperature would be $950^{\circ}C$, the total heat transfer from the fire to the upper part of the vessel 134 kW/m^2, re-radiation to the environment 127 kW/m^2 and the internal radiation loss to the liquid layer 7 kW/m^2. In this situation, the temperature drop through the wall required to conduct heat to the inner surface would be less than $1^{\circ}C$, justifying the assumption of negligible temperature difference. Estimating a heat transfer coefficient for natural convection from the wall to the vapour gives h = 4 W/m^2. $^{\circ}C$; for the prevailing temperature differences of about $500^{\circ}C$ this mechanism gives a heat transfer rate of 2 kW/m^2 which justifies the assumption of negligible convective heat transfer. However, from 8 minutes, when strong boiling occurs, forced convection from the walls to the steam has a larger effect. The observed drop in wall temperature of about $70^{\circ}C$ over a period of 2 minutes corresponds to a heat loss of 11.3 kW/m^2 to the steam and a transfer coefficient of 16 W/m^2. $^{\circ}C$.

The rate of heat transfer to the water up to the onset of boiling is estimated as 154 kW. On terminating the trial at 22 minutes, 100 litres of water remained giving a mean transfer rate of 260 kW over the period 8.5-22 minutes. This represents an average transfer rate of 219 kW to the water over the period 0-22 minutes. With 200 litres of water in the tank, the non-wetted surface area is 2.37 m^2 and the wetted area is 1.63 m^2. The non-wetted surfaces radiate to the liquid at 7 kW/m^2, as discussed above, making a contribution of 17 kW. The balance of 202 kW is therefore received from the wetted surfaces at an average rate of 124 kW/m^2. This is equal to the heat received at the outer surface, since radiation from the low temperature wetted areas may be neglected, and agrees well with the estimate of 134 kW/m^2 derived above for the non-wetted surfaces. Thus a value of Q = 130 kW/m^2 is representative of the average heat transfer rate from the fire to the vessel.

Trials with propane

Ten trials were conducted with propane, one on a non-insulated tank and nine with different insulation systems applied to the tank; details of the systems will not be given here but they are regarded as representative of the current state of the art for reasonably practicable systems. Types of system tested included active systems based on epoxy resins that intumesce under fire engulfment conditions, passive systems based on mineral fibres and combinations of active and passive systems.

One significant point that affected the assessment of the trials data was the variable performance of the pressure relief valves, all of which were preset to open at 1.7MPa (250 psi) but which in practice opened in the range 0.7-1.8MPa (100-265 psi). This is thought to be due to either weakening of the spring or damage to the valve seating by the effects of the fire. A further significant point was the range of ambient temperatures for the trials - from $-11^{\circ}C$ to $+20^{\circ}C$ - corresponding to starting pressures ranging from 0.35 MPa (50 psi) to 0.95 MPa (140 psi). The time to first operation of the relief valve is therefore an unreliable indicator of the effectiveness of the insulation system.

In the majority of trials, the boiling of the propane was in the film boiling regime as evidenced by the typical wall temperature traces in Fig 5. In contrast to Fig 3 there is little difference between the upper and lower surface temperatures, indicating that the liquid propane is not wetting the walls but is supported on a vapour film.

Furthermore, the lowest internal thermocouple (not shown) gave an erratic reading due to intermittent exposure to heated propane vapour. The temperature difference between the walls and the liquid is 220°C at 20 minutes, characteristic of film boiling.

In contrast the data shown in Fig 6 shows nucleate boiling of propane in trial conditions. This regime persists until about 75 minutes when there is a transition to film boiling, possibly caused by a spontaneous instability in the regime or, more probably, by a breakdown in the insulation causing a large increase in heat transfer. Further analyses are being made of the nucleate boiling/film boiling transitions in these trials.

The significance of the occurrence of film boiling is a) that the walls are vulnerable to thermal weakening due to the lack of cooling by liquid propane and b) that the heat transfer to the liquid is not a reliable indicator of the transmission characteristics of the insulation because the ratio of heat gain of the walls to that of the liquid is different for the two boiling regimes.

The heat transmission rates are calculated from the wall temperature and vapour pressure measurements in each trial. The trial period is divided into two parts - from ignition to the first operation of the relief valve (during which the propane mass is constant) and from this first operation to the completion of discharge of propane. During the latter period it is assumed that the propane discharge rate is constant. The transmission rates to the propane and to the system (container and propane) are summarised in the tables below. Each of the insulation systems reduces the heat transfer rate, when compared with the rate with no insulation present. A further criterion of insulation performance is the ability to keep wall temperatures below some temperature related to the loss in mechanical properties of the steel. In the US Federal Register test for insulation systems for fire engulfment conditions (2) the pass/fail criterion is the time at which a temperature of 427°C (800°F) is achieved by an insulated plate heated in a prescribed way. Times to achieve this temperature are recorded in Table 2 for each trial, as a basis for comparison.

Discussion

When formulating a view on the technical requirements for a thermal insulation system for the protection of vessels containing hazardous liquids, it is important to have defined objectives. One can, in principle, add more and more insulation to achieve better and better protection but the law of diminishing returns would make such a policy uneconomic.

Vessels in need of protection fall broadly into two classes - mobile and fixed. A mobile installation, such as a road or rail tanker or portable cylinder, creates different risks from a fixed installation. The involvement of a road tanker in a crash could apply considerable mechanical forces to an insulation system, calling for a strength specification, and in an overturning accident, where damage to the pressure relief valve might occur or operation of the pressure relief valve could send a jet of flame horizontally for tens of metres, the paramount need might be to delay the operation of the pressure relief valve for a period sufficient to allow evacuation. On the other hand, at a fixed installation the pressure relief valve should be sited so that its operation does not create a further hazard and the need might be to discharge the vessel contents safely without the vessel walls overheating.

These requirements are not incompatible but they may call for different numerical criteria. As a first approximation, the time to a critical wall temperature is independent of scale, filling ratio and nature of liquid (over a reasonable range) because in the worst case heat transfer through the insulation is heating a dry wall which is losing heat internally mainly by radiation, which is not a very efficient process, judging by the low value of F obtained in the trials with water.

On the other hand, the time to first operation of the pressure relief valve depends on the size of the vessel, the filling ratio and the nature of the liquid and to predict this time one needs to consider a specific situation and to have information on the heat transfer/time characteristics of the insulation under fire engulfment conditions as well as data on the internal heat transfer and boiling processes.

In the US Federal Register Test referred to above (2), two test procedures are specified - a simulated pool fire test of 100 minutes duration and a torch fire test of 30 minutes duration. Both tests are calibrated by exposing a bare steel plate on one face only to a hydrocarbon flame; in the simulated pool fire test, a plate 12 x 12 x ^{5}/8 inches is required to attain $427^{O}C$ in 13 ± 1 minutes exposure while in the torch fire test a plate 48 x 48 x ^{5}/8 inches is required to attain $427^{O}C$ in 4.0 ± 0.5 minutes exposure. In each case, the plate is free to radiate to a cold environment from the face not exposed to flame. The temperature variation of the plate is given by equation (2) for these heating conditions, with F = plate emissivity. Equation (2) shows that, all other things being equal, the time to a given temperature is proportional to plate thickness.

As can be seen from Fig 3, the 5mm thick wall achieved a temperature of $427^{O}C$ in approximately 1.6 minutes from ignition; as stated earlier, the fire takes approximately 30 seconds to develop so that the effective heating time to $427^{O}C$ is 1.1 minutes. On the above basis, the experimental arrangement described in this paper would therefore heat 16mm (^{5}/8 inch) plate to $427^{O}C$ in about 4 minutes and hence heating severity is comparable to the US Torch Fire test although it achieves this by a somewhat lower value of heat input to the heated face and a lower heat loss from the unheated face; it can maintain this heating environment for up to 100 minutes.

The pass/fail criterion for the US tests is that with an insulated plate, the plate temperature shall not exceed $427^{O}C$ during the period of the test. Accepting that the fire engulfment trials test insulation materials with a severity comparable to the Torch Fire test, this would imply that insulation systems with a value greater than 30 minutes in the final column of Table 2 would stand a good chance of passing the US Test.

Rankings of insulation performance can be made from the data in Table 2. On most counts material 'I' has the highest ranking and material 'C' the lowest but there is less consistency with the intermediate materials. Some materials perform markedly better before PPV lift than afterwards (they deteriorate with time as the active material is consumed) while others are more consistent with time. Whether

boiling is in the film or nucleate regime affects the heat transfer to
the liquid. Some care is therefore required in the definition of test
criteria.

The fire engulfment trial procedure is judged to be of adequate
severity for its intended purpose. It has merit as a procedure in that
it tests the complete system in a realistic way but it is expensive to
conduct. Further work is in hand on a test analogous to the US Torch
Fire test but on a smaller scale which could be used as a screening
test.

TABLE I: Heat Inputs to Liquid Contents of Tanks

INSULATION	LIQUID	HEAT INPUT TO LIQUID (kW)		
		TO PRV LIFT	LIFT TO END	ALL
NONE	WATER	154*	260	219
NONE	LPG	123	113	114
A	LPG	10.9	18.3	16.1
B	LPG	10.9	34.4	18.3
C	LPG	16.5	22.9	21.3
D	LPG	3.2	18.3	8.5
E	LPG	8.3	14.5	12.2
F	LPG	3.6	8.0	7.7
G	LPG	5.7	10.8	9.9
H	LPG	1.9	4.4[+]	2.8[+]
I	LPG	0.8	1.3[+]	0.9[+]

TABLE 2: Heat Inputs to the the System of Tank and Contents

INSULATION	HEAT INPUT TO SYSTEM (kW)			MINUTES TO 427°C
	TO PRV LIFT	LIFT TO END	ALL	
NONE	227*	265	250	2
NONE	315	278	283	3.5
A	35.0	42.8	40.0	20
B	32.0	70.4	49.0	15
C	60.0	57.4	56.0	8
D	17.0	38.3	24.0	52
E	60.0	20.8	36.0	13
F	25.0	20.6	22.0	13
G	14.0	14.8	15.0	82
H	8.3	9.3+	8.6+	55
I	2.2	4.2+	2.8+	{100

* To onset of boiling when test liquid was water
+ estimated values

Acknowledgements: The work described in this paper was carried out on
a contract basis with Shell Research Ltd and the valuable input of Dr
I Williams is hereby acknowledged.

References:

(1) Wright, J. M., Fryer, K. C. "Alternative fire protection systems for LPG Vessels". GASTECH 81. Conf. on LNG & LPG. Hamburg, Germany 1981

(2) US Code of Federal Regulations, Title 49, 179.105-4, 405-407 1981

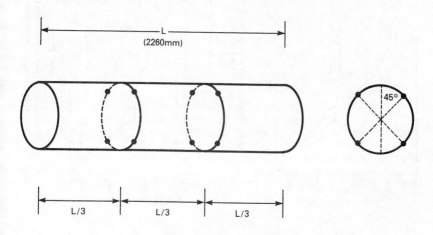

Fig. 1 - Positions of wall thermocouples

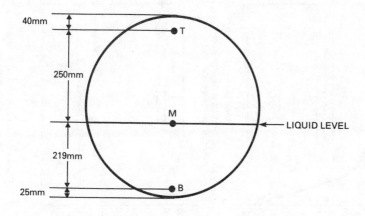

Fig. 2 - Positions of internal thermocouples

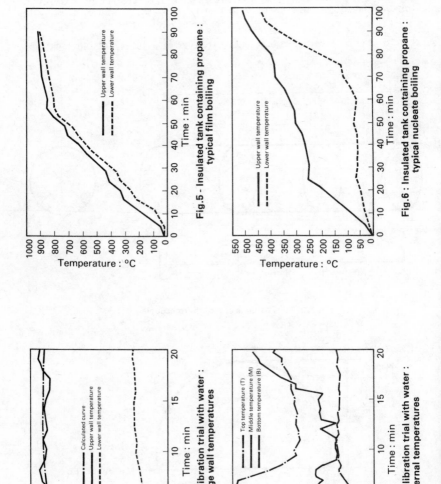

Fig.5 - Insulated tank containing propane : typical film boiling

Fig.6 : Insulated tank containing propane : typical nucleate boiling

Fig.3 - Calibration trial with water : average wall temperatures

Fig.4 - Calibration trial with water : internal temperatures

EXPLOSIONS IN FLAMMABLE SMOKES FROM SMOULDERING FIRES

K N PALMER

FIRE RESEARCH STATION, BOREHAMWOOD, HERTS.

SUMMARY

Many combustible solid products of the chemical and allied industries are able to undergo smouldering combustion, generating smoke and flammable vapours which can lead to explosion. The phenomenon has received little detailed study but a number of explosions in buildings and plant units have been attributed to the buildup of smoke and its explosion by a subsequent ignition, sometimes by the smouldering fire itself.

Experimental investigations have been made using natural and synthetic polymeric foams, which smoulder readily from small sources of ignition, and the combustion products included flammable gases and vapours, together with a flammable liquid aerosol suspension. As combustion proceeds in a closed vessel, the oxygen concentration diminishes as the concentration of flammable atmosphere builds up, until ultimately the oxygen level can be sufficiently reduced to prevent the propagation of an explosion flame in it. However there can be a range of atmospheres in which the concentrations of flammable materials and of oxygen are both sufficiently high for an explosion flame to propagate and cause a pressure rise.

Consideration is given to the means of detecting a hazardous atmosphere, but detectors conventionally used for smoke or flammable vapours may not respond to explosible smoke atmospheres. Alternative means of detection may therefore be necessary.

INTRODUCTION

Explosions in which flammable gases, dusts, or mists, are mixed with air and ignited are recognised as an important hazard to life and property. The rapid movement of flame in an explosion prevents the use of emergency fire fighting methods, whether automatic or manual, and in buildings results in people having insufficient time to use the means of escape. The rapid rate of combustion generates a correspondingly large volume of hot products which, if unable to escape by venting, will raise the pressure within the plant unit or building. The rise in pressure may be sufficient for structural damage to be caused because much plant and buildings in the process and allied industries are incapable of withstanding pressures of a fraction of 1 bar without damage resulting. A further consequence in buildings is that fire protection measures may be nullified, either because automatic sprinklers or fire detectors are damaged before they have time to operate effectively.

In these situations the flammable atmosphere is present before ignition occurs. An alternative situation is that combustible may be present, either solid or solid impregnated with liquid, before ignition occurs.

The ignition leads to a fire, which is often subdued, and which produces flammable combustion products. In the open air, or in a well ventilated enclosure, the combustion products would be removed and diluted rapidly but in an enclosure such as a closed plant unit or building the products may accumulate. These products may either be gaseous or aerosol. Because ventilation is low or non-existent the combustion will consume oxygen from the air originally present at the time of ignition, as well as generate flammable products. Conditions may develop in which the accumulation of flammable products is sufficiently rapid for a flammable atmosphere to form before the oxygen concentration falls to a level which cannot support flame propagation. As a source of ignition is already present, namely the fire, a hazardous situation exists in that an explosion may take place. Because the atmosphere is somewhat vitiated the maximum explosion pressure and the maximum rate of pressure rise may not be as severe as with explosions in air but hazardous emission of flame and generation of pressure are possible.

The experiments described in this paper were exploratory in nature to ascertain the conditions under which such explosions might occur, to obtain some measure of the vigour of the explosion, and to obtain information on the response of smoke detectors to this type of fire situation.

Alternatively, combustion products may continue to accumulate and the oxygen concentration is depressed below the level at which flame propagation can occur. Smouldering of the fuel may continue even though flaming is unable to persist. However if the enclosure is subsequently opened, by a door or a hatch, then air can enter, the oxygen concentration can rise and the conditions for a flammable mixture to form may then be present. The access of air may also cause the smouldering combustion to transform into flaming, which then ignites the atmosphere causing an explosion. Such a sequence is hazardous for operatives or fire fighters and frequently leads to their injury.

PREVIOUS INCIDENTS

A comprehensive survey of reported incidents has been made by Croft[1] who was able to show that two types of fire could lead to explosion, which he called smouldering and developing fires respectively. The characteristics of a smouldering fire were that the smoke had accumulated over a relatively long time, because the rate of growth of the fire was slow, and hence the fire atmosphere was relatively cool. This enabled personnel to enter the atmosphere to search for the seat of the fire and they were thus vulnerable to being in the atmosphere when an explosion occurred. The other type was described as a developing fire in which the rate of growth was relatively fast, so that the atmosphere of the fire was hot, combustible surfaces present, if not actually burning, would be heated and generating combustible vapours and if conditions were favourable the rate of flame propagation through the vapours was fast enough to be designated as explosion. In addition, secondary explosions can occur from the presence of flammable liquids which generate vapour by evaporation and contribute to the flammable atmosphere. By its nature, a developing fire is likely to involve combustion with flame, and is outside the scope of the present paper, which is concerned with smouldering fires and the longer timescale.

In a reported total of 52 smouldering fires which led to explosions, Croft showed that 13, ie 25 per cent, were caused by chemicals or associated products such as plastics and rubbers. The remaining materials causing explosion were cellulosic, such as textiles, paper and wood products, and may be regarded as contents of buildings or packaging materials. Thus as far as the process industries are concerned the problem can arise both in the manufacture and storage of the product itself, or at a later stage when it is packaged.

The casualty rate in the fires was high. In the 52 fires there were 116 casualties, including 30 fatal. The average was 2.23 casualties per fire, and the ratio of non-fatal to fatal casualties was 2.87. Taking the UK as a whole, the number of casualties per fire in the chemical and allied industries is usually about 0.09 and the ratio of non-fatal to fatal casualties is about 27. These figures are for the year 1980 in which the total number of fires in the industries, attended by public fire brigades, was 649. Smouldering fires which lead to explosions are clearly much more hazardous to life than are the more usual straight-forward fires.

INVESTIGATION OF EXPLOSION INVOLVING FOAMED RUBBER

The background to a specific incident was studied in detail by Woolley and Ames[2]. The incident involved foamed rubber slabs, stored as a stack of mattresses, situated in a building of volume 176m^2 with all doors and windows to the exterior closed, except for one window partly open. The alarm was given when smoke was seen to be emerging from the window and the fire brigade, wearing breathing apparatus, entered the store on arrival but were unable to locate any flames even though a thorough search was made. The smoke was cool and dense and while the search was being carried out windows were opened to ventilate the smoke. At this point an explosion occurred with flames emerging through the open windows and the door. The explosion was reported to have been suffic-iently severe to break some of the windows but did not damage the main structure of the building. The flash of the explosion originated in the general area near the stack, and it was followed almost immediately by an intense fire. There were 6 casualties in the incident, two of them fatal.

Subsequent laboratory investigations showed that the foamed rubber was readily ignited to smouldering by a glowing source such as cigarette, or to flaming by a small flame. The propagation of smouldering in the foamed rubber was accompanied by the generation of visible smoke which, if collected as it was being formed, would burn quietly with a stable flame. If allowed to continue over a period, within an enclosure, 1.4m^3 volume, the flammable smoke steadily accumulated and at 30 minutes visi-bility within the compartment was of the order of 0.1m. Insertion of a small flame ignited the aerosol within the compartment causing an explosion, which vented itself through a weak plastic diaphragm covering the front of the enclosure. At the time of ignition, the oxygen con-centration within the enclosure was 19.3 per cent by volume. It was concluded that in the incident a smouldering fire had been initiated in the foamed rubber and the resulting aerosol had caused the explosion, the source of ignition being the fire itself.

Separate tests with a block of the foamed rubber in air showed that smouldering could be initiated at a point on its surface and would propagate radially at a velocity of 0.55cm per minute in a horizontal direction and slightly slower vertically downwards. This smouldering rate is relatively rapid for still air conditions, but when account is taken of the density of the foamed rubber the rate is comparable with that of other organic solids, as shown in Table 1, based on results reported elsewhere[3,4].

TABLE 1
Smouldering of organic materials

Material	Particle diameter μ m	Smouldering rate cm/min	Density g/cm^3	Smouldering rate x density g/cm^2.min
Foamed rubber latex	–	0.55	0.065	0.036
Beech wood dust	190	0.096	0.30	0.029
Cork dust	< 65	0.30	0.18	0.054
Fibre insu-lating board	–	0.12	0.25	0.030
Grass dust	<65	0.24	0.28	0.067
Coal dust	<104	0.10	0.60	0.060

CHARACTERISTICS OF SMOULDERING

From Table 1 it can be seen that the product of smouldering rate and density, whilst not constant, varies over a narrow range as compared with the smouldering rate itself or the effect of an airflow on the smoulder-ing rate[2]. As smouldering is essentially a process controlled by diffusion of air, the rate of smouldering would be expected to vary inversely with the density of fuel, and to be affected by ash formation, liquefaction, or volume change at the smouldering zone. An exact proportionality between smouldering rate and density would thus not be expected, since different fuels have different non-combustible contents and physical properties whilst burning.

However, as Croft showed, many organic materials, whether cellulosic or not, can produce smokes which may cause explosions. From Table 1 it can be seen that the rate of generation of such smokes is likely to be dependent upon the product of smouldering rate and density, and not highly dependent upon the type of combustible involved. Croft was not able to make a systematic study of the timescale in the various incidents, but it appears that usually the smouldering must have been in progress for one or more hours in order that an explosible atmosphere could accumulate. The smouldering rates listed in Table 1 support this evidence, since they are sufficiently slow for burning to be required for at least an hour for the accumulation of sufficient smoke to cause a damaging explosion. For example, if smouldering should be initiated at a point on the surface of the combustible and it then propagates at a constant rate, according to Table 1, and after one hour a hemisphere will have been consumed which had an original mass of several kilogrammes. Only a fraction of the mass would have been converted to flammable smoke but a kilogram of aliphatic hydrocarbon can form more than 20m^3 of flammable mixture in air. This volume of flammable mixture, if ignited, could cause an explosion and structural damage in plant units such as ovens or dryers, or in compartments of buildings.

The means whereby flammable atmospheres can be formed, from smouldering materials, are straightforward and do not require highly specific circumstances.

TESTS FOR COMBUSTIBLE PRODUCTS

There are no standard tests currently available for measuring the ability of a material to develop a flammable smoke, from smouldering combustion. This lack of a test does create difficulty in predicting which materials, when involved in a fire, could give rise to the hazard. Some tests are available in Germany for measuring the explosion properties of carbonisation products in small furnaces or other vessels[2]. However the combustible vapours are produced by heating powdered material rather than allowing it to smoulder, so that the composition of the vapours may well be different.

The apparatus for measuring the minimum ignition temperature of carbonisation products is shown in Fig. 1 and consists of a horizontal cylindrical furnace containing a concentric test chamber. One end of the chamber is constructed with a narrow diameter inlet so that powder can be blown in from a squeeze bulb whilst the other end is a flap which is hinged and is deflected by the pressure produced should ignition occur. The diameter of the test chamber is 75 mm and its length, including the drawn down portion, is 190 mm. It contains a circular concave deflector whose leading edges are about 50 mm from the flap and parallel to the walls of the chamber. The temperature of the walls is controlled by means of a thermocouple and thermostat, and the temperature of the deflector is also measured.

The method of using the apparatus is to heat the furnace to its maximum temperature, switch off the heating and wait until the temperature of the deflector is higher than that of the wall of the chamber. The powder is then blown into the furnace and observation made, using a mirror, as to whether a flame appears. The furnace is cleaned by a blast of compressed air and the test repeated at lower temperatures until no ignition occurs. In some cases there is delay of several seconds before ignition occurs, because time is required for flammable vapours to be produced from the powder introduced.

Another test is available which measures the explosion pressure of carbonisation vapours in a 1.5 litre spherical vessel. The vapours are again generated by heating, rather than by smouldering, and are allowed to mix with air within the sphere before ignition.

Neither test is specifically designed for flammable smokes from smouldering fires, and both tests can only be applied to powders.

DETECTION OF FIRE

The selection of automatic detectors for flammable smokes can cause
problems. Fire detectors are commercially available which respond to
smoke, or heat or flame. In the present context detection of flame is
clearly inappropriate, whether the detector responds to the infrared or
ultraviolet radiation, and detectors which respond to heat, that is to a
function of temperature rise, are also likely to be unsatisfactory since
as has already been stated the flammable smoke atmosphere may be cool.
Detectors which respond to smoke operate using either optical or ionisation methods.

Optical detectors for smoke rely on the obscuration or the scattering of
light by the smoke particles, or they combine both effects. A light
beam is focussed on or near a light-sensitive cell and where smoke is
present the output from the cell changes, and this change is utilised
to give an alarm. Provision has to be made to prevent variation in
ambient light conditions from affecting the system, by having the system
in a light-tight container or by modulating the light source. Generation of smoke or of change in refractive index of the air due to heat
emission from the fire causes the performance of the beam to change and
an alarm to be given. Optical detectors would be suitable for fires in
which cool flammable smokes were generated, and the detectors should give
an alarm long before the volume being protected has accumulated sufficient
smoke to generate a flammable atmosphere throughout the volume.

Ionisation detectors for smoke have an ionisation chamber as the sensor.
The ionisation chamber contains two electrodes, across which a potential
difference is maintained, and a radioactive source, usually an alpha-
particle source such as americium. The source, by ionising the air,
produces positive and negative gas ions which travel to the electrodes
of opposite polarity to themselves; this movement generates an electric
current. When smoke particles enter the chamber, the charged ions
attach themselves to some of the particles. As these charged particles
are heavier than the gas ions, they move more slowly between the electrodes and have more time than the gas ions to meet particles or ions
of the opposite sign and be neutralised. The result is a reduction in
the current flowing in the chamber. This reduction, the amount of which
depends upon the number of smoke particles in the chamber and their
diameter, is used to actuate an alarm by means of a suitable electric
circuit. The response of the detector clearly depends upon the physics
of the particles including their mass, and ability to take up charge and
to coalesce. For this reason the sensitivity of ionisation detectors
to smokes from various types of fire can vary widely. Smokes which
are of small particle size are readily detected. Such smokes can arise
from fires with flame, and smouldering sources exposed to the atmosphere
particularly if they are close to the detector. Where the smoke from
the smouldering has to pass through unburnt fuel, or through combustion
residues, then there is greater opportunity for coalescence and for
other processes associated with the ageing of the smoke to take place.
Ionisation detectors are less sensitive to smokes which have aged, and
some recent work has been reported studying this effect[6].

FLAMMABLE SMOKES

A detailed investigation has been carried out in which the explosibility of smoke atmospheres has been studied[6]. Only a summary is presented here. The apparatus consisted of an explosion chamber with open front, of dimensions 1.1 x 1.1 x 0.7m deep. The chamber was instrumented so that the variation of oxygen concentration and temperature with time could be measured, using a paramagnetic oxygen analyser preceded by filters to remove aerosols, at a sampling rate of 1 litre/min, and a commercial sheathed chromel-alumel thermocouple (1.5mm external diameter) respectively. An ionisation fire detector was also placed in the chamber so that its time of operation could be noted in respect of the state of smouldering. It was a standard commercially available detector modified to produce a continuous analogue voltage output, and which had previously been calibrated in a smoke test tunnel. The triggering point of the detector was 78 millivolts.

Tests were carried out with both foamed rubber and flexible polyurethane foam, ignited by smouldering sources.

The foamed rubber was in the form of two pieces which were placed on top of each other to form a block 0.5 x 0.5 x 0.2m high, inside the enclosure. Smouldering was initiated at the centre of the top surface of the block, using a glowing cigarette, and the front of the enclosure was closed by means of a lightweight plastic film. The results, summarised in Fig.2, showed that the ionisation detector operated approximately 14 minutes after initiation of the smouldering, at which time the temperature of the smoke had risen only about 4K above ambient, and the oxygen concentration of the atmosphere had decreased from 20.8 to 19.8 per cent by volume. The density of the foamed rubber was 0.072 g/cm^3. At the time that the ionisation detector operated the smoke concentration was dense to the eye and it would have been helpful if the alarm had been given earlier, to facilitate protection measures. Comparable results were found with polyurethane foam and, indeed, in one case the smoke concentration was sufficiently high for it to be ignitable by a flame, before the detector operated.

It was concluded that, in general, for this type of smouldering combustion, that detectors operating on the optical principle would give an earlier alarm than the ionisation detectors.

A more detailed study of the generation of flammable atmospheres from the smouldering of foamed rubber was carried out by Tonkin and Croft[7]. The volume of their explosion chamber was 1.41m^3, the dimensions being 1.5 x 1.5 x 0.75m deep. The front of the enclosure was closed by a transparent plastic film.

The enclosure was equipped with 7 gas sampling points, for monitoring of the oxygen concentration, as well as chromel-alumel thermocouples for temperature measurement. No smoke detectors were included. Two blocks of foamed rubber, each 480 x 380 x 100mm were placed one on top of the other on a metal tray supported by bricks inside the enclosure. The density of the foamed rubber was 0.064 g/cm^3. Smouldering was initiated by means of a 20 swg Kanthal wire laid between the two rubber blocks near to their centres. The experiment was started at the time that the wire was energised, and the current continued until smouldering had been established within the rubber blocks. This was indicated by the

discolouration of the upper surface of the top block and the appearance
of smoke above it.

Flammability of the smoke was assessed by inserting a small propane flame
into the upper part of the enclosure through an inlet socket, and by
making observations to whether or not propagation of flame occurred in
the smoke.   Propagation was usually accompanied by bursting of the
plastic film.   The temperature in the enclosure, before ignition, ranged
between ambient and 100°C.   The maximum explosion pressure was sometimes
as high as 7 kN/m² even though the enclosure was generously vented.   A
smaller vent, such as would be expected in a plant unit or building would
have resulted in higher pressures.

In some of the experiments the smouldering rubber was weighed so that a
progressive record was obtained of the weight loss.   By repeating the
experiments it was possible to establish a range of oxygen concentrations
within the enclosure within which sufficient smoke had accumulated to
cause a flammable atmosphere but which was not so vitiated that flame
propagation could not occur.   A specimen of the results is shown in
Table 2 which indicates that between oxygen concentrations in the range
18 to 20.5 per cent by volume, a flammable atmosphere could be generated.
If the burning were allowed to continue until the oxygen concentration
had decreased below 18 per cent then introduction of the flame did not
cause an explosion.   However if air were subsequently introduced into
the box, to partly dilute the smoke mixture then flame propagation could
occur and sometimes develop spontaneously, the smouldering material acting
as an ignition source for the flammable atmosphere.   This situation would
in practice represent the disturbance of an atmosphere by the opening of
a door or window, causing the introduction of air and simultaneously
providing an ignition source, so that explosion developed.

Table 2

Flammability of smoke

Oxygen concentration % by volume	Result of application of flame
20.8	No ignition
20.3	No ignition
19.8	Ignition
19.6	Ignition
18.5	Ignition
17.9	No ignition
17.2	No ignition
16.4	No ignition

From measurement of the rate of loss of mass an estimate was made of the
linear smouldering rate of the foamed rubber, as follows.   The mass $\triangle$
M after time t, if the linear smouldering rate is S and the density of
the foamed rubber is $\mathcal{C}$ is given by

$$\triangle M = \frac{4}{3}\pi_{\mathcal{C}}\,(St)^3 \qquad\qquad (1)$$

A plot of $\Delta M^{\frac{1}{3}}$ against t is given in Fig. 3 and shows, at least in the early part of the burning, that the relationship was linear. From the gradient of the line the value calculated for the smouldering rate S was 0.62 cm/min, which compares with direct measurement in a similar sample of foamed rubber of 0.55 cm/min (Table 1).

Further consideration shows that the consumption of the foamed rubber was an inefficient combustion process. The stoichiometric mixture of rubber (assumed C_5H_8) in air is 87 g/m^3 so there was originally sufficient air in the box to consume 87 x 1.41 = 123 g of foamed rubber. The mass loss in the experiment was between 900 g and 1 Kg, at which time only about half the oxygen initially in the enclosure had been removed. Thus most of the loss of mass arose because of vaporisation from the foamed rubber, rather than combustion, so it is not remarkable that an atmosphere sufficiently flammable to cause a smoke explosion could readily be generated.

Finally, some consideration was given to the removal of flammable smoke atmospheres in practical situations should the necessity arise either in a building or in a plant unit. The problem is in two parts, a recognition of the existence of the hazardous atmosphere and implementation of methods for its safe removal. As has been already stated, the use of smoke detectors may enable the existence of the burning to be discovered, but the smoke detector itself does not determine whether or not the atmosphere is explosible. Equipment designed for use in the detection of flammable gas atmospheres is generally vulnerable to interference by aerosols, or may not respond to a flammable aerosol. Reliable use of flammable gas detectors cannot therefore be assumed. The problem is not yet solved but there are now prototype flammable atmosphere detectors which may in due course enable the problem to be readily solved.

Tonkin and Croft carried out experiments in an enclosure of moderate volume to study the methods whereby a flammable atmosphere could be removed. Three possible methods were studied

1. Inerting the enclosure atmosphere with nitrogen
2. The use of water sprays
3. The use of high expansion fire fighting foam

Details of the results will be reported elsewhere[7], but on the scale investigated the use of inert gas appeared to give the most promising results.

CONCLUSIONS

A range of common organic solids has been shown to give rise to flammable smokes when allowed to smoulder in an enclosure.

Commercially available smoke detectors can give early warning of the burning, but not all detectors are sufficiently sensitive.

Although oxygen in the enclosure has to be consumed in order to support the smouldering which generates the flammable smoke, sufficient may remain for flame propagation to occur after an appropriate concentration of smoke has accumulated.

Instruments for the detection of flammable smoke atmospheres have not yet been developed commercially, for general application.

Flammable smoke atmospheres can be removed from buildings or plant units by several methods, and possibly the introduction of inert gas is the most effective.

ACKNOWLEDGEMENTS

This paper forms part of the work of the Fire Research Station, Building Research Establishment, Department of the Environment. It is contributed by permission of the Director, BRE.

REFERENCES

1. Croft W M.   Fire Safety Journal $\underline{3}$ (1980/81) 3 - 24.

2. Woolley W D and Ames S A.   The explosion risk of stored foam rubber. Building Research Establishment Current Paper 36/75 April 1975.

3. Palmer K N.   Combustion and Flame $\underline{1}$ (1957) 129 - 154.

4. Cohen L and Luft N W.   Fuel Lond. $\underline{34}$ (1955) 154 - 163.

5. Leuschke G and Osswald R.   VDI-Berichte 304 (1978) 29 - 38.

6. Kennedy R H and Rogers S P.   (In publication)

7. Tonkin P S and Croft W M.   (To be published)

Fig 1 Sketch of furnace for measuring minimum ignition temperatures of powders

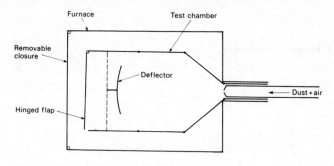

Fig 2 Response of fire detector to smoke from foamed rubber

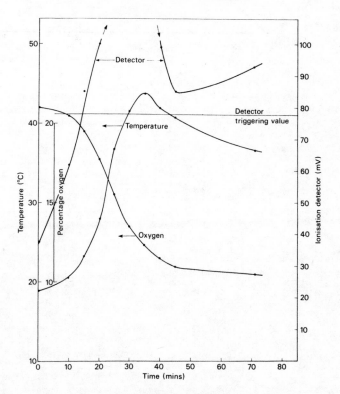

Fig 3   Application of equation (1)

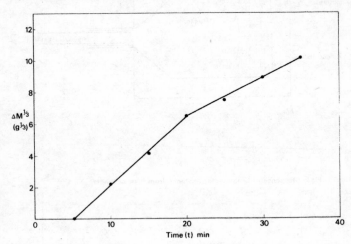

An Experimental and Theoretical Study of Blast
Effects on Simple Structures (Cantilevers)

D. K. Pritchard

Explosion and Flame Laboratory, Buxton, Derbyshire, England

INTRODUCTION

Over the last decade the consequences of unconfined vapour cloud explosions (UVCE's) at or near petro-chemical complexes and nuclear power
plants have figured prominently in the safety considerations of these
sites. The destructive power of this type of explosion is illustrated by
the widely publicised incidents that have occurred at Beek, Flixborough
and Pernis (references 1 and 2). Not surprisingly a lot of effort, both
experimental and theoretical, has gone into developing methods for
predicting the effects and investigating the mechanisms of UVCE's.
Theory predicts that high flame velocities (100 m s^{-1} or greater) are
necessary to produce the overpressures required to account for the
observed damage and that, at similar overpressures the rise-times and
positive phase durations are longer in the blast wave from a UVCE than in
the blast wave from high explosives. These differences affect the damage
caused by the blast wave. To date no convincing experimental simulation
of a UVCE has been made nor has any adequate explanation of flame
acceleration effects been given.

The major sources of data on the overpressure-time histories of UVCE's
are actual accidents. The damage caused to pieces of equipment or
structural elements can be used diagnostically for determining the
magnitude and possibly the duration of the overpressure at given
locations. In the past the most has not been made of this diagnostic
information for various reasons, for example because of confidentiality
considerations, or lack of knowledge of damage mechanisms.

A problem with the diagnostic approach is that of relating damage to
overpressure or impulse (the pressure/time integral). There is a
considerable body of information on the effects of blast from high
explosives and nuclear bombs (though not all published in the open
literature) that may be used to estimate overpressures. However, due to
the different characteristics of the UVCE blast wave such data may lead
to errors in estimates of overpressure and impulse (reference 3). A
blast wave of long duration but low peak pressure could cause more damage
to certain objects, depending on their frequency of response, than a
blast wave of higher peak value but shorter duration. It is for these
reasons that the practice of assessing UVCE's in terms of "TNT
equivalent" has been so widely criticised. There is therefore a need for
a more detailed analysis of the diagnostic methods used on UVCE
accidents.

A programme to investigate the effect of UVCE type blast waves on simple
structures has been undertaken at the Health and Safety Executive's
Explosion and Flame Laboratory at Buxton, as an aid to accident investigation. In this paper experimental results are presented for solid
rectangular and tubular cantilevers and comparisons made with predicted

tip displacements using simple elastic-plastic theory. The theory is also of value in its own right since it can also be used as an input in the design of geometrically similar structures, for example flare stacks, distillation columns, to withstand expected blast loadings in order to prevent further loss of flammable inventories.

EXPERIMENTAL TECHNIQUE

For the experimental work a steel explosion gallery (69 m long by 1.22 m diameter) was used. The blast waves were generated by igniting mixtures of methane/air, at the closed end of the gallery. The test structures were placed in the middle section of the gallery (37 m from the closed end). Although this arrangement is a confined system it produced explosions which gave, in the test section, peak pressures of up to 90 kPa, rise-times in the range 20 to 100 ms and positive durations of about 200 ms, values believed to be representative of UVCE blast waves.

Flame travel and explosion pressures (equivalent to side-on over-pressures) were monitored with photocells and strain gauge type and piezo-electric pressure transducers placed inbye and outbye of the test structure. The pressure transducers had been calibrated dynamically by suddenly applying a known pressure. Dynamic pressures in the test section were measured by one of two methods. The first method used a stagnation gauge, which was mounted within the gallery cross-section. The difference between the stagnation pressure and the side-on over-pressure (measured at the gallery wall) is essentially equal to the dynamic pressure for the sub-sonic particle flows produced by the explosions. The stagnation gauge was calibrated by the same method as used for the pressure transducers. The second method used a reed anemometer, which consisted of a mild steel reed clamped at one end with a half bridge foil type strain gauge arrangement bonded to the base of the reed. The anemometer was calibrated by hanging weights from the tip of the reed. As the natural frequency of the reed (370 Hz) is much greater than the frequency of the explosion loading (about 2 Hz) the dynamic loading from the blast and the static loading from the calibration were assumed to be equivalent (see chapter 2 of reference 4). Pressures measured by the anemometer were drag pressures on a rectangular cross-section: to convert to dynamic pressures they were divided by the appropriate drag coefficient.

Agreement between the dynamic pressures obtained by the two methods was within 10% or less in most cases. Of the two methods the stagnation gauge method is subject to greater error as it involves taking the difference between two numbers of similar magnitude. In order to obtain the drag pressure and hence the loading on the structure, the measured dynamic pressure was multiplied by the appropriate drag coefficient. This step is a potential weak link in the determination of loading on the test structure. Most of the drag coefficient data in the literature has been obtained under steady flow conditions for rigid structures, and its application to the dynamic conditions of a structure deforming in a changing air flow is open to question. However, agreement with theory for deformations in the elastic regime indicate errors introduced are not large.

The test section in the gallery was equipped with glass windows which allowed cine records to be taken of the test structures undergoing deformation. In most experiments a foil type strain gauge was bonded to the inbye face of the test cantilever. To protect the gauges from damage they were coated with a layer of silicone rubber. Gauge outputs were calibrated using the built in calibration system of the conditioner/amplifier.

A limitation of using an explosion gallery is that oscillations are set up in the gallery air column by pressure wave reflections from the closed and open ends. The reflected rarefaction from the open end of the gallery results in a double peak in the dynamic pressure-time profile, the inward travelling rarefaction wave producing an outflow of gas. This is illustrated in Figure 1 which shows a typical dynamic pressure profile. It is because of these oscillations that comparisons between theory and experiment have been limited to maximum tip displacements, as they always occurred before the first reflected pressure wave reached the test section.

To avoid boundary layer effects and non-uniform loading, the cantilevers were mounted well clear of the gallery walls. Cantilever lengths were limited to 0.75 m and in most experiments were considerably less than this value. Tests with miniature pressure transducers mounted in a rigid structure indicated that the pressure field, at a given time, was uniform over the gallery cross-section used for the cantilever work.

EXPERIMENTAL RESULTS

Experimental results have been obtained for aluminium and mild steel solid rectangular cantilevers and mild steel tubular cantilevers. A hot rolled mild steel and aluminium 2014 were used for the rectangular cantilevers. This particular grade of aluminium was chosen because of its low sensitivity to strain rate effects. The tubular cantilevers were made from an annealed cold rolled mild steel. Cantilever dimensions were chosen in order to give permanent deflections in the range 20 to 90°. Estimates indicated that rectangular cantilevers with a thickness of 3 mm and tubular cantilevers of 12 mm outside diameter and 0.8 mm wall thickness would give the desired deflections. Accordingly cantilevers were cut from flat bars with nominal cross-sections of 3 x 25 mm (lengths ranging from 0.4 to 0.45 m) and 20 swg (0.92 mm) or 22 swg (0.71 mm) 12 mm diameter tube (lengths in the range 0.5 to 0.75 m). The cantilevers were used as cut from the bar or tube without any further conditioning or machining.

Yield stresses of the cantilever materials were determined from samples cut at regular intervals from the bars and tubes. In the case of samples that showed no definite yield stress the 0.2% proof stress was used. These measured yield stresses were corrected for strain rate effects, induced by the finite rate of loading used in the determinations, by applying the Cowper-Symonds relationship.

$$\frac{\sigma_T}{\sigma_S} = 1 + \left(\frac{\dot{\varepsilon}}{D}\right)^{\frac{1}{P}} \qquad (1)$$

As an exact elastic theory exists then agreement between observed and predicted elastic tip deflections would be strong evidence for no serious errors in the experimental procedure and measured drag pressures. Therefore, a few experiments were carried out with 6 mm thick aluminium and mild steel cantilevers in order to obtain a purely elastic response. Table 1 presents the results for maximum tip deflections and measured strains.

COMPARISON WITH THEORY

Elastic Deflections

The theory for predicting the elastic deflections of a uniformly loaded cantilever has been obtained by Jones (reference 5) by applying modal analysis to the forced vibration of a cantilever beam. The basis of the modal method is that each normal mode of vibration can be treated as an independent one-degree system and then superimposed to obtain the total response.

The equation of motion for the nth mode, Lagrange's equation, is

$$\ddot{A}_n(t) + \omega_n^2 A_n(t) = \frac{f(t) \int_0^L q(x) \phi_n(x) dx}{m \int_0^L \phi_n^2(x) dx} \qquad (2)$$

from which can be obtained A_{nst} the modal static deflection. Modal response is given by

$$A_n(t) = A_{nst} (DLF)_n \qquad (3)$$

and total response by

$$y(x,t) = \sum^n A_{nst} (DLF)_n \phi_n(x) \qquad (4)$$

Analytical solutions for tip deflections, for different loading functions, can be obtained from equation (4) and expressions for stress, bending moment, rate of curvature, etc, by differentiating these solutions. In applying the equations it was found to be sufficient to sum over the first three modes, inclusion of further modes making negligible difference to the calculated displacements.

In the initial comparisons a bi-linear, ie triangular, approximation to the loading history (obtained from the drag pressure-time history) seemed a reasonable approximation to make (see Figure 1). To avoid the complications caused by the pressure pulsing of the gallery, comparisons with experiment were not carried beyond the time to the end of the first peak of the drag pressure-time history. However, predicted maximum tip displacements were appreciably lower than the observed maximum and the shape of the predicted tip displacement-time histories were a poor match to the observed. Closer examination of the loading functions suggested that in the majority of experiments a tetra-linear or tri-linear loading function, would be a better approximation. This function takes into account the precursor and/or broad peak exhibited in many of the drag pressure profiles. The comparisons between the observed and predicted maximum tip displacements, using the best approximation to the loading history, are shown in Table 1. Comparisons between the measured and predicted strains are also included in the Table 1. The shape of the predicted tip displacement-time profiles were also now in good agreement with the observed profiles. A representative example of this agreement is shown in Figure 2. On average theory underpredicted the measured displacements and strains by 10% for aluminium cantilevers and 17% for mild steel cantilevers. The difference between the two materials may be significant, for example, the elastic theory used will not account for differences in material dynamic effects. It should be noted that in experiment 165 (Table 1) the observed strain exceed the static yield strain ($4900\mu\epsilon$) and in experiment 161 it was very close to this value, even though the values of q_1/q_s were less than 0.67. This means that in the former experiment and possibly in the latter the outermost fibres at the root of the cantilever would have undergone plastic deformation, giving rise to an increased tip displacement.

Elastic-Plastic Deflections

To model the maximum deflections of a uniformly loaded cantilever under-going permanent deformation it is necessary to include the elastic and plastic deflections. The elastic-plastic model used is based on a method developed by Symonds (references 5 and 6), in which it is assumed the cantilever response is either elastic or plastic, and occurs in distinct separate stages. The elastic stage ends when the global yield condition (the bending moment at the cantilever root equals the static collapse moment) is reached. Velocity at the beginning of the plastic stage is matched to that at the end of the elastic stage by minimising the kinetic energy difference between the two stages.

The elastic theory described above was used to calculate the tip deflections up to the global yield condition and a rigid-perfectly plastic theory from yield to maximum deflection. For the rigid-perfectly plastic model the basic equations, neglecting rotary inertia, are

$$\frac{\partial Q}{\partial x} = m\ddot{y} - q \qquad (5)$$

$$Q + \frac{\partial M}{\partial x} = 0 \qquad (6)$$

Combining equations (5) and (6) and integrating gives an expression for $\ddot{y}$ from which can be derived expressions for tip displacement and velocity.

Trial comparisons indicated that further modifications to this simple theory were required. First, it was necessary to include strain rate effects, especially for the mild steel cantilevers. This was achieved by using the dynamic collapse moment (M_T) for the yield criterion. For example for rectangular cantilevers M_T is evaluated from the following expression (derived from equation (1)).

$$M_T = M_s \left[1 + \frac{2P}{1 + 2P} \left(\frac{\dot{X} H}{2D} \right)^{\frac{1}{P}} \right] \qquad (7)$$

using the values of $\dot{\kappa}$ calculated in the elastic stage. Secondly, as a cantilever bends the loading acting on it depends not only on the drag pressure-time history, but also on the angle of deflection. For large deflections the reduction in loading can be appreciable, thus a correction to the loading was included in the plastic stage calculations but not in the elastic stage. Thirdly, in the case of aluminium cantilevers the tip deflections at the global yield point were large enough for the assumption of linear elastic behaviour to be invalid. It was, therefore necessary to include a correction for non-linear effects in the elastic calculations. In the present experimental work the response was almost quasi-static and therefore the published graphs (see reference 7) of the ratio of linear to non-linear deflections, for static loading, were used to correct the calculated elastic tip deflections and velocities.

The elastic-plastic theory described here was initially developed for solid rectangular cantilevers, but it will also apply to tubes provided there is no change in the cross-section during deformation. Comparisons with experiment were, therefore, limited to cases where a post-test examination revealed no significant change in the tube cross-section at the point of bending.

Comparisons between experiment and theory are shown in Figures 3, 4 and 5. Results are plotted according to the ratio q_1/q_s for ease of presentation but this does not fully represent the different loadings as it takes no account of the duration and thus the differing dynamic load factors. For this reason experiments with higher q_1/q_s ratios can give lower tip deflections and for some experiments (see Figure 4) plastic deformations are observed and predicted for q_1/q_s ratios less than one. In Figures 3 and 5, for mild steel, the higher value for the predicted deflections were obtained ignoring strain effects and lower value obtained by including strain rate effects. The inclusion of the non-linear correction to the elastic stage made negligible difference to these values. In the case of aluminium, Figure 4, the strain rate correction was negligible and was therefore ignored, the upper value being obtained by using linear elastic theory and the lower value by applying the non-linear correction.

Overall, for all three types of cantilever, the agreement between experiment and lower value predicted by theory is fair. However, there is the same tendency in all three cases for theory to underpredict for the lower q_1/q_s ratios and overpredict for the higher ratios. There are several possible reasons for this behaviour:

i)   The coefficients used in the Cowper-Symonds equation were typical values for a mild steel and a strain rate insensitive aluminium, thus the time coefficients may be slightly different for the actual materials used in the tests.

ii) No account was taken of the possible reduction in the drag coefficient, and thus loading on the cantilever, for large deflections.

iii) Though the stress-strain diagrams obtained for the cantilever materials indicated strain hardening is minimal its effect on the maximum tip deflection may be significant for large deflections.

iv) In the theory it is assumed that the cantilever responds elastically up to the global yield point and plastically above this point, while in practice there is a short period, just before the global yield point, when the response is elastic-plastic.

CONCLUSIONS

1   The good agreement between experiment and theory for elastic deflections validates the experimental technique.

2   Fair agreement has been obtained between the predictions of the elastic-plastic theory and the observed maximum tip deflections. However, further development of the theory is required as well as additional comparisons with experiment, over a wider range of loading conditions and cantilever dimensions.

3   More information is required on the dynamic material properties of metals and the aerodynamic characteristics of simple structures as they deform under a dynamic loading.

ACKNOWLEDGEMENT

The contribution of Prof. N Jones of the Department of Mechanical Engineering, University of Liverpool, to this work in developing the elastic and elastic-plastic theories and his advice on experimental procedures, is gratefully acknowledged.

SYMBOLS USED

$A(t)$	—	Modal amplitude
D	—	Strain rate constant
DLF	—	Dynamic load factor
$f(t)$	—	Loading time function
H	—	Cantilever thickness
L	—	Cantilever length
m	—	Mass per unit length
M	—	Bending moment
p	—	Strain rate constant
q	—	Load per unit length
Q	—	Transverse shear force
x	—	Distance from cantilever root
y	—	Transverse deflection
$\dot{\varepsilon}$	—	Strain rate
$\phi(x)$	—	Characteristic shape
$\dot{\kappa}$	—	Curvature rate
σ	—	Yield stress
ω	—	Natural frequency

Subscripts —   n — nth mode
          S — Static value
          T — Dynamic value
          1 — Peak value

REFERENCES

1 Gugan, K, (1979). Unconfined Vapour Cloud Explosions Institute of Chemical Engineers, George Godwin Ltd, London.

2 Sadee, C, Samuels, D E, and O'Brien, T P, (1976). The characteristics of the explosion of cyclohexane at the Nypro (UK) Flixborough Plant on 1 June 1974. Journal of Occupational Accidents 1(3), 203-235.

3 Roberts, A F and Pritchard, D K, (1982). Blast Effect from Unconfined Vapour Cloud Explosions. Journal of Occupational Accidents 3, 231-247.

4 Biggs, J M, (1964). Introduction to Structural Dynamics. McGraw-Hill, New York.

5 Jones, N. Private communication.

6 Symonds, P S, (1980). Finite elastic and plastic deformations of pulse loaded structures by an extended mode technique. International Journal of Mechanical Science, 22 (10), 597-605.

7 Wang, T M, (1969). Non-linear bending of beams with uniformly distributed loads. International Journal of Non-Linear Mechanics. 4, 389-395.

TABLE 1 - Elastic results, experiment and theory

Experiment No.	q_1/q_s	Max.tip deflection : mm		Strain : $\mu\epsilon$	
		Observed	Predicted	Observed	Predicted
(a) Mild Steel Cantilevers					
157	0.52	26.2	21.9	–	–
159	0.52	23.3	21.3	1495	1310
158	0.66	44.8	29.7	2680	1760
167	0.66	52.3	47.7	2310	2195
83	0.69	46.5	37.3	–	–
(b) Aluminium Cantilevers					
168	0.27	34.9	32.9	2780	2460
163	0.33	34.9	32.8	2730	2460
165	0.45	78.5	66.1	6040	4850
161	0.60	66.9	64.5	4860	4815

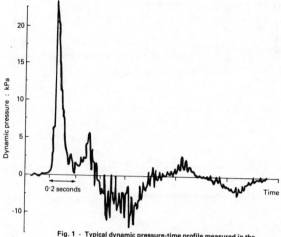

Fig. 1 - Typical dynamic pressure-time profile measured in the test section with the reed anemometer

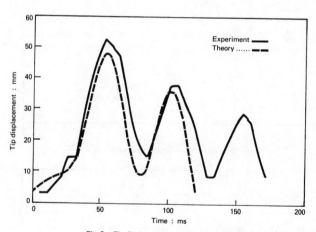

Fig. 2 - Tip displacement history for experiment 167

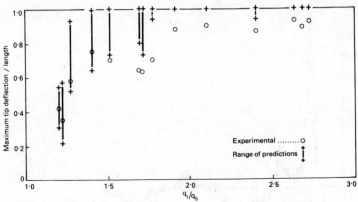

Fig. 3 - Mild steel solid rectangular cantilevers

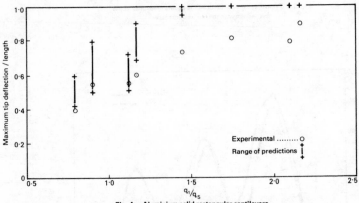

Fig. 4 - Aluminium solid rectangular cantilevers

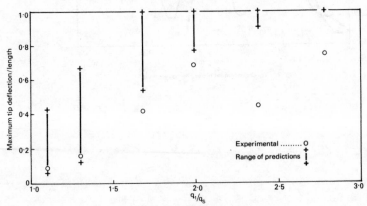

Fig. 5 - Mild steel tubular cantilevers

Explosion Hazards

Chairman

G. Ferraiolo
*University of Genoa,
Italy*

PURGE LIMITS FOR PARAFFIN HYDROCARBONS IN AIR AND NITROGEN

P. ROBERTS* AND D.B. SMITH

LONDON RESEARCH STATION, BRITISH GAS CORPORATION. (*CURRENTLY AT
RESOURCES AND EXTERNAL AFFAIRS.)

The flammability limits of fuel-air-nitrogen
systems have been re-examined for four paraffin
fuels, methane to butane. Measurements were made
in a spherical closed vessel (0.6 m diameter) with
central ignition. Flame propagation was monitored
by visual observation and by measuring pressure
changes. Arguments are presented in favour of this
method for determining limits rather than the
traditional tube apparatus.

Particular attention was paid to purge limits.
These were shown to be significantly wider than
currently accepted values. It is recommended that
the new values supersede earlier data as the basis
for practical purging operations. Continued use
of these earlier data could lead to hazards during
purging since the potential exists for the formation of mixtures capable of supporting vigorous
combustion with significant overpressures.

For operations involving purging by dilution, the
new limits impose additional purge gas requirements of between 5 and 20%, depending on the fuel.

INTRODUCTION

The purging of vessels holding flammable gases is a standard operation
and procedures exist for the safe undertaking of such tasks (1). An
important pre-requisite to the prescribing of sound procedures is a
proper knowledge of purging end-points. Purge limits for most commonly
occurring fuels were apparently established long ago, with results
subsequently collated and analysed in standard data sources (1, 2).
However, while investigating the effects of combustion in near-limit
mixtures, we were suprised to discover supposedly limit mixtures burning
quite vigorously. Because of this, we decided to re-measure the purge
limits, initially for methane but later extended to ethane, propane and
butane.

Traditionally flammability limits* have been measured by observing flame
propagation along a tube. Currently accepted purge limits were largely
determined in this way. It has been long acknowledged that wall

---

*In this paper, flammability limit is taken to refer to the boundary
between flammable and non-flammable mixtures at any part of the
fuel-air-inert system. Purge limits refer more specifically to the tip
of the flammability envelope, where inert concentrations have their
maximum values.

quenching effects will interfere if the tube is not sufficiently wide. A tube diameter of 50 mm was thought sufficient by Coward and Jones (3) and this has been commonly used. More recent work (4, 5) has shown that this is incorrect and that wall effects are still significant. This must shed some doubt on the validity of tube experiments. In recent years, there has been a growing trend towards studying flammability in spherical vessels to remove wall effects (6, 7, 8). This work has thrown new light on flammability characteristics. But two potential problems arise. First, if the vessel is not large enough to allow reasonable flame travel, the flame may not be well removed from influence of the ignition source. The 3.65 m diameter spherical vessel used at the Bureau of Mines (6,7) is ideal. Some other òr'· has involved vessels which are too small for the true picture to emerge. Secondly, the flammable mixture is totally confined. Since no venting occurs, pressures build up during combustion. This may lead to wrong conclusions being drawn about flammability at constant pressure. But it should produce a relevant picture of events that could arise during purging and provide safe guidance for practical operations.

EXPERIMENTAL

Flammability limits were measured in a closed spherical vessel (0.6 m in diameter) with central spark ignition. Flame propagation was monitored by visual observation and by pressure change measurement. Mixtures were made by measuring partial pressures; the composition of some mixtures was checked by gas chromatography and by paramagnetic oxygen analysis. All tests were conducted at initial conditions of 1 bar and ambient temperature (ca 20°C). Further details of apparatus and procedure have been given previously (9).

RESULTS

General Observations

Near-limit combustion in spherical closed vessels shows several forms, depending on the type of fuel and the mixture stoichiometry. These various forms of burning have been observed and described before (8, 9). We have designated these different types of combustion by symbols as shown in Table 1.

It should be noted that not all four types are observed at all limits. For example, consider the case of fuel-air mixtures (with no added inert). Methane shows all four types on the lean side but only the last two on the rich side. With the other fuels studied, the position is precisely reversed, with four at the rich and two at the lean limit. On addition of inert, the patterns on the lean and rich sides necessarily have to merge. The detailed results show where this merging occurs.

The first type of burning (denoted ↑) is gentle and consumes only a small fraction of the mixture. Therefore it produces only small pressure rises in the vessel. On the other hand, the final type (denoted by ✲) is vigorous and propogates rapidly throughout the mixture; accompanying pressure rises are considerable. For all fuels studied, the over-pressures obtained fell into fairly distinct bands, depending on the type of combustion. An indication of the overpressures obtained is shown in Table 1.

TABLE 1   TYPES OF FLAME PROPAGATION OBSERVED IN THE CLOSED
           SPHERICAL VESSEL

Brief description of flame behaviour	Symbol	ΔP (bar)
Cap of Flame went up to roof and died out.	↑	0 - 0.2
Cap of flame went up to roof, then started to burn back down, dying out in the top half of the vessel.	↕	0.2 - 0.6
Flame went up to roof; then burned back down to the bottom, consuming all the mixture	↕	1.4 - 2.3
Flame burned outwards in all directions from the spark	✛	2.3 - 3.0

Presentation of Data

Flammability limit data for the fuel-air-N_2 systems are presented on
triangular diagrams, as shown schematically in figure 1. Purge limits
are then defined by the two tangents XA and YF.

When purging out of service (fuel to inert operations), the operation
starts at apex F (100% fuel). On adding purge gas, the composition
follows triangle edge FI. If the purge is stopped at any point and air
allowed to enter, then the composition follows a line towards apex A.
Point X represents the purge limit, since for any composition beyond X,
addition of air cannot produce a flammable mixture.

Similarly, when purging into service (air to inert operations), the
operation starts at A (100% air), and moves towards I as inert is added.
The tangent YF determines the purge limit Y.

Although triangular diagrams are a good means of displaying the data, in
practice we used an alternative means of graphical representation, as
described by Zabetakis (2). This alternative procedure leads to better
precision in locating the purge limits.

Detailed Results

Data for the four fuels methane to butane are displayed graphically in
Figures 2 to 5. The extreme limit is of most concern for purging
considerations, but the limits for the other forms of combustion are also
given, because of their implications for purging operations, as discussed
later. For clarity, the two forms of partial combustion (↑ and ↕) are
combined on the diagrams. (Note that in these diagrams the fuel and
inert axes cover the range 0 to 50% and air 50 to 100%). It can be seen
that the general shape of the flammability envelope is similar for the

four fuels, though there are significant differences of detail, particularly in the way the different types of burning are manifest.

The data are displayed in tabulated form in Tables 2 and 3. Purge limits appropriate for purging out of service (fuel to inert operations) are shown in Table 2. The major concern is with the outer limit of combustion, and this is indicated as the main limit. But limits for the other forms of burning are also shown. Data are represented in two ways: (i) as the ratio of inert to combustible (I/C) as used by Zabetakis (2) and others, and (ii) as the percentage fuel in the fuel + inert mixture. The latter is probably more useful, since the monitoring of the fuel concentration is the most practical means of deciding when a purging end-point has been reached, although the former is particularly suited to le Chatelier calculations for mixed fuels.

For purging into service (air to inert operations), the appropriate limits are shown in Table 3. Again main and subsidiary limits are given. Data are displayed in two ways : (i) as percentage of inert in the inert + air mixture, as used by Zabetakis (2), and (ii) as the percentage oxygen in the limit mixture. The latter is generally of more practical use.

DISCUSSION

Validity of the Present Work

Before discussing the implications of this work for purging operations, we seek to establish that the procedures used and data obtained are valid.

The outer limits measured in this work depend on an initial upward propagating flame which is essentially free from container effects. The flame has no contact with the walls (until it reaches the roof). Because the proportion of the mixture consumed during this initial upward burning phase is very small, pressure rises are small and should have little effect on the measurements. Therefore we believe this method of determining flammability limits is sound and that our data are reliable.

For limits in air (no inert), we found fairly good agreement with the accepted values given by Zabetakis (2). This comparison was reported previously (9). Where differences did arise, our limits were generally somewhat wider. We consider that heat losses and wall quenching in tube experiments are probably the major causes of the discrepancies. Of course it may be argued that our experiments did not allow sufficient flame travel and that in a larger vessel, flames in mixtures at the extreme limits might extinguish before reaching the wall. This would explain any tendency for our limits to be wider. But we find excellent agreement with the results for butane obtained in a 3.65 m diameter vessel (6), suggesting that the use of a larger vessel would make very little difference.

As a final comment on techniques, we note that traditional tube experiments cannot provide the amount of detail about flammability behaviour that is obtained from experiments in spherical vessels.

Comparison with Previous Data on Purge Limits

Although our fuel in air limits agree fairly well with accepted data,

TABLE 2   PURGE LIMITS : PURGING OUT OF SERVICE

Fuel	Purge Limits: Upper figure = % Fuel (Lower figure) = I/C ratio			
	Main Limit	Subsidiary Limits		
	↑	↨	↓	⊕
Methane	11.4 (7.8)	12.7 (6.9)	14.9 (5.7)	21.7 (3.6)
Ethane	6.0 (15.8)	6.1 (15.5)	6.8 (13.8)	11.2 (7.9)
Propane	5.1 (18.7)	5.2 (18.3)	5.4 (17.5)	9.0 (10.1)
Butane	4.2 (22.6)	4.3 (22.2)	4.4 (22.0)	6.9 (13.4)

TABLE 3   PURGE LIMITS : PURGING INTO SERVICE

Fuel	Purge Limits: Upper figure = % O (Lower figure) = % N in N + Air			
	Main Limit	Subsidiary Limits		
	↑	↨	↓	⊕
Methane	12.0 (43)	12.4 (41)	13.1 (37.5)	15.2 (27.5)
Ethane	10.3 (51)	10.5 (50)	11.4 (45.5)	14.3 (32)
Propane	10.8 (48.5)	11.4 (45.5)	12.0 (43)	14.3 (32)
Butane	10.9 (48)	11.7 (44.5)	12.1 (42.5)	14.4 (31.5)

this is not the case for purge limits. In Table 4 we have compared our data with those given by Zabetakis (2), and by Coward and Jones (3).

It can be seen that our limits are significantly wider than those recommended in either of the two standard sources. Our data demand purging to lower concentrations of fuel or oxygen. We suggest there are several reasons why the earlier data are wrong. The accepted values often rely on single determinations of purge limits. To compound this, much of the original work is now very old (some of it over 60 years). Although the basic methods adopted were the same as in later work, the procedures were often quite different: for example with gas supplies (no cylinders) and gas handling techniques. The tubes in which the tests were conducted were sometimes of small diameter, thus exaggerating wall quenching effects. (In contrast to this, the limits in air alone (no inert) have been more thoroughly and more recently measured. This probably explains the better agreement observed here.)

Implications for Practical Purging Operations

The differences between the present work and accepted data have important consequences for practical purging operations. The significance of these differences is most easily demonstrated by reference to Figures 2-5, on which the Zabetakis and Coward and Jones limits are marked. For purging out of service, the earlier limits would, on the addition of air, produce mixtures well inside the flammable region. This is particularly true for propane and butane, where mixtures inside the zone of complete combustion (↕) could be formed. Such mixtures would burn quite vigorously and yield overpressures as high as 2 bar. The effect of not attaining the limit tends to be more marked in the fuels ethane to butane than in methane because of the structure of the flammability zones. With these higher hydrocarbons, lean combustion is dominated by the more severe forms of burning. With methane, on the other hand, lean mixtures show all types of combustion with the gentler forms much in evidence.

For purging into service, the position is similar though somewhat less marked. For methane, ethane and propane, the previous limits cross the flammable region close to the boundary of complete combustion (↕) yielding potential overpressures in the range 0.8-1.5 bar. For butane, this zone is well-penetrated; overpressures could now be as high as 2 bar.

Of course when practical purging operations are considered, it is customary to impose an additional safety factor (typically 20 or 30% of the limit). It might be argued that this lessens the force of the discussion above and renders safe an operation using the old purge limits. But the safety factor is introduced to allow for any errors in determining the end-point in the practical operation or for possible non-uniformity in purging. Its purpose should not be to compensate for any flaws in the basic data.

In any event, purge end-points using currently accepted data with a 20% safety factor, are barely adequate. The worst case involves the purging out of service of a vessel containing propane. The Zabetakis limit with the safety factor is 14.7 x 0.8 = 11.8% fuel. This lies inside our limit (11.4%), and is clearly a potential hazard.

Purge Requirement

Adoption of the purge limits reported in this paper has a secondary implication for practical purging operations, since the choice of purge end-point often dictates the amount of purge gas required and the length of time for purging. For some operations, particularly those involving pipelines, conditions close to plug flow can be achieved. In this case, purge gas requirement approximately equals the volume to be purged, irrespective of the purge limit adopted. Indeed, the whole operation is essentially unaffected by the choice of purging end-point. At the other extreme is the use of purging by dilution which occurs when tank gas and incoming purge gas become well-mixed. Such conditions may arise when purging large tanks or gas-holders. The tank gas concentration falls

TABLE 4   COMPARISON OF PURGE LIMITS

Fuel	Reference		
	This work	Zabetakis	Coward and Jones
Purging out of Service		(% Fuel)	
Methane	11.4	14.7	13.9
Ethane	6.0	7.2	7.2
Propane	5.1	6.0	6.2
Butane	4.2	5.6	5.5
Purging into Service		(% Oxygen)	
Methane	12.0	12.8	13.0
Ethane	10.3	11.7	11.4
Propane	10.8	12.0	12.0
Butane	10.9	12.6	12.4

exponentially, according to the expression:

$$C = C_0 \exp (-u.t/V)$$

where C is the tank gas concentration at some time t; C_0 is the original concentration; u is the flow rate of purge gas, and V the tank volume to be purged. The quantity u.t/V is useful, since it is equal to the ratio (volume of purge gas)/(volume of tank). With C equal to the purge end-point, it gives the purge requirement in terms of tank volume.

Selected values of the purge requirement calculated in this way are presented in Table 5 for the four fuels, for purging into and out of service. Two sets of data were chosen for this : the present limits and Zabetakis data, both with a 20% safety factor. The increases in purge requirements in going from the Zabetakis limits to ours are also given in the table. These increases vary considerably (from 5 to 20%). The worst case is the purging into service of butane vessels, where 20% extra purge is required.

CONCLUSIONS

1. New data are presented on the flamability limits of fuel-air-nitrogen systems, for the four fuels methane, ethane, propane and

butane. Our limits are consistently wider than currently accepted values, including the limits for purging.

2. We recommend that the purge limits given in Tables 2 and 3 be used as the basis for establishing purging procedures for these fuels. Use of current limits allows the formation of mixtures capable of sustaining vigorous combustion with potential overpressures of up to 2 bar. Even with a 20% safety factor, current data are barely adequate.

3. When "purging by dilution", the use of the new limits will impose an additional penalty in terms of purge gas requirements. This varies between 5 and 20% for the different fuels.

ACKNOWLEDGEMENT

The authors wish to thank British Gas for permission to publish this paper.

REFERENCES

1. American Gas Association, 1975, "Purging Principles and Practice".

2. Zabetakis, M.G., 1965. "Flammability Characteristics of Combustible Gases and Vapours", U.S. Bureau of Mines, Bulletin No. 627.

3. Coward, H.F., and Jones G.W., 1952, "Limits of Flmmability of Gases and Vapours", U.S. Bureau of Mines, Bulletin No. 503.

4. Andrews, G.E., and Bradley, D., 1973, "Limits of Flammability and Natural Convection for Methane - Air Mixtures", 14th Symposium (Int.) on Combustion, p 1119, The Combustion Institute, Pittsburgh, Pa.

5. Jarosinski, J. and Strehlow R.A., 1978. "The Thermal Structure of a Methane - Air Flame Propagating in a Square flammability Tube", Aeronautical and Astronautical Engineering Department, University of Illinois, Technical Report AAE 78 - 6.

6. Furno, A.L., Cook E.B., Kuchta J.M., and Burgess D.S., 1971, "Some Observations on Near-Limit Flames", 13th Symposium (Int.) on Combustion, p 593, The Combustion Institute, Pittsburgh, PA.

7. Sapko, M.J., Furno A.L., and Kuchta, J.M., 1976, "Flame and Pressure Development of Large-Scale CH_4 - Air - N_2 Explosions", U.S. Bureau of Mines, Report of Investigation No. 8176.

8. Crescitelli, S., Russo, G., Tufano V., Napolitano F. and Tranchino L., 1977, Comb.Sci.Techn., 15, 201.

9. Roberts, P., Smith D.B. and Ward D.R., 1980, "Flammability of Paraffins in Confined and Unconfined Conditions", I.Chem.E.Symp.Ser., 58, 157.

Symposium Series No. 82

TABLE 5   PURGE GAS REQUIREMENTS

Fuel	Purge requirement/Vessel Volume This work	Zabetakis	Increase %
Purging out of service		
Methane	2.39	2.14	11.7
Ethane	3.04	2.85	6.5
Propane	3.20	3.04	5.3
Butane	3.39	3.11	9.1
Purging into service		
Methane	0.78	0.72	8.4
Ethane	0.94	0.81	16.0
Propane	0.89	0.78	13.7
Butane	0.88	0.73	20.0

DIAGRAMS

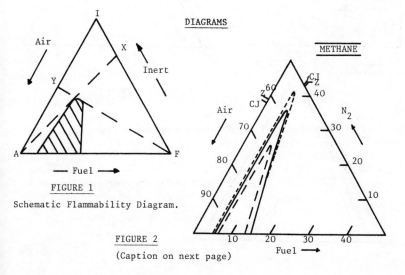

FIGURE 1
Schematic Flammability Diagram.

FIGURE 2
(Caption on next page)

E9

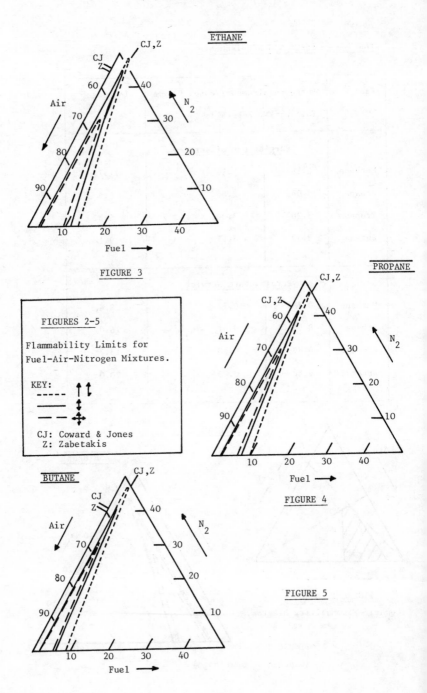

ETHANE

FIGURE 3

FIGURES 2-5

Flammability Limits for
Fuel-Air-Nitrogen Mixtures.

KEY:

CJ: Coward & Jones
Z: Zabetakis

PROPANE

FIGURE 4

BUTANE

FIGURE 5

THE EFFECT OF VENTILATION ON THE ACCUMULATION AND DISPERSAL OF HAZARDOUS GASES.

M.R. Marshall

BRITISH GAS, MIDLANDS RESEARCH STATION, SOLIHULL, WEST MIDLANDS.

INTRODUCTION

Whenever gases and vapours are contained in pipework and process plant, the possibility exists that an accidental release of these substances can occur. In addition, if the primary containment is itself located within a confining enclosure (for example, a building) the leaking gas or vapour could accumulate and produce a potentially hazardous situation. For example, if an uncontrolled release involves a flammable gas or vapour, an explosion and/or fire hazard could exist.

Whilst good plant design and construction, complemented by planned maintenance and well-defined operating procedures, should minimise the likelihood of an accidental release this cannot be guaranteed. Means of preventing such an escape subsequently producing a hazardous situation therefore should always be considered. An effective measure is to provide sufficient ventilation so that any credible accidental leakage can be dispersed safely, i.e. the maximum concentration of gas or vapour is prevented from exceeding the safe limit.

In most cases, the ventilation air will be provided by natural ventilation (wind driven and/or buoyancy driven) and therefore the air change rate will be variable, depending on weather conditions. However, there are general design principles on which to base such ventilation systems.

Whilst there is guidance on the ventilation requirements of buildings, for example incorporated in Standards, Building Regulations, Codes of Practice, etc., this is usually concerned with air change rates to ensure comfort, adequate combustion air supplies and, in some cases, flue product dilution. Hence, it is not necessarily relevant to the dispersal of leakages of potentially hazardous gases and vapours. Similarly, few studies on the mixing behaviour of leaking gases relevant to an industrial situation have been reported in the literature[1].

This paper presents the results of a study into the way leaking gas mixes with air and the modifying role of ventilation on this process. The application of the data to the design of effective natural ventilation systems for buildings is discussed together with the limitations inherent in such systems.

EXPERIMENTAL ARRANGEMENT

The four most important factors which influence the manner of build up of gas concentration from a release into an enclosure are the density of the gas released, the nature of the gas leakage source, ventilation and the enclosure volume. The effect on the mixing process of these parameters has been investigated in an experimental study. The main series of tests was carried out in a cubical enclosure of $20.6m^3$ volume. Additional tests were made in a smaller $8m^3$ rectangular shaped enclosure and also in buildings (20.6-$55.6m^3$). Although the experimental programme was designed to determine the behaviour of leaking gas under naturally ventilated conditions, for ease of experimentation and to provide control of the variables, the main experimental study was carried out using controlled air flow conditions. However, comparisons with the behaviour under these essentially forced ventilation conditions were obtained from experiments in naturally ventilated buildings.

Unventilated Situation

For the tests carried out under zero ventilation conditions, the variables investigated were gas specific gravity, gas flowrate, gas leak velocity, leak position and orientation, duration of leakage and enclosure volume. Table 1 shows the range of these variables.

Table 1: Range of Variables in Experiments

Variable	Unventilated	Ventilated
Gas specific gravity	0.46 (town gas) 0.6 (natural gas) 1.5 (propane)	0.6 (natural gas)
Gas flow rate, Q_g	0.48 to 9.75	0.28 to 9.75
Gas velocity, v	0.3 to 122	0.7 to 61
Leak position, z	0 to 1.0	0.04 to 0.95
Leak orientation	Upwards, horizontal, downwards	Upwards, horizontal
Test Volume, V	8 (tall chamber) 20.6 (cubic chamber) 20.6 to 55.6 (test buildings)	20.6 (cubic chamber) 20.6 to 55.6 (test buildings)
Air flow rate, Qa	-	5.1 to 122
Ventilation patterns	-	Upwards, cross flow, downwards.

N.B. Units are defined in the Nomenclature at the end of the paper.

Ventilated Situation

The tests carried out were designed to determine how the mixing behaviour observed under zero ventilation conditions is modified by the presence of ventilation air. A wide range of ventilation regimes (i.e. flow patterns and air change rates) has been studied. These include not only the basic upward air flow ventilation regime but also the reverse, downward flow condition and an evaluation of the influence of the position of the inlet and outlet ventilators under cross flow conditions. The ranges of the variables studied are shown in Table 1, Column 3.

DISCUSSION OF RESULTS

In this and subsequent sections, it is to be assumed that the discussion refers to a release of buoyant gas unless otherwise stated. However, it is considered that the same general conclusions can be drawn for dense gases, other conditions being similar, except that the total system is inverted (i.e. a buoyant gas release under an upward air flow regime is expected to behave similarly to a dense gas leakage with downward air flow).

Unventilated Situation

When a gas is released into an enclosure, it will mix with the available air due to the actions of turbulent jet mixing, buoyancy and turbulent interaction with air supplied by ventilation. Gas mixing by molecular diffusion alone is extremely slow in comparison to these other effects and in most practical cases can be ignored. For many situations, the part of the enclosure volume in which turbulent jet mixing is the dominant process is likely to be restricted to a region close to the leakage position. Thus, in an unventilated situation where there is no turbulent mixing produced by ventilation air, the density of the gas released will have a dominant influence on the way in which gas–air mixture accumulates. Consequently, leakages of dense gases, such as propane and butane and flammable vapours such as petrol, would be expected to result in the formation of gas-rich layers near the floor; conversely, buoyant gases such as natural gas would be expected to form layers near the ceiling. This idealised behaviour is illustrated in Figure 1. The results of the experiments carried out under zero ventilation conditions confirm this pattern of mixing and indicate that a

(a) Dense gas forming a floor layer   (b) Buoyant gas forming a ceiling layer

Figure 1. CONCENTRATION PROFILES IN AN UNVENTILATED ROOM
UNIFORM LAYERS OF BUOYANT AND DENSE GASES

well defined layer is formed between the level of the leak and the ceiling (Figure 2). With a leak near to the ceiling, a shallow layer of high concentration is formed, whereas an identical leak nearer the floor fills the enclosure volume between the leak position and the ceiling with a uniform mixture of lower concentration. The higher gas concentration produced by a high level leak, evident in Figure 2, is a consequence of a smaller fraction of the total volume of air in the enclosure being involved in the mixing process.

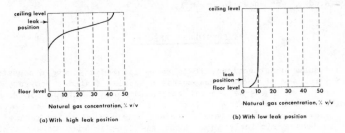

Figure. 2. TYPICAL EXPERIMENTAL NATURAL GAS CONCENTRATION
PROFILES IN AN UNVENTILATED ROOM

An important result of these experiments (as demonstrated in Figure 2) is that, for leaks which occur near the floor, the time taken for uniform conditions to be established within the layer is short. This implies that the time scale over which a gas release can be considered to behave as either a turbulent jet or a buoyant plume is very short. Consequently, it is a misconception to assume (as suggested by Leach and Bloomfield[1]) that, following the onset of leakage, a layer of high concentration is formed at the ceiling which then gradually increases in depth with time. The experimental data show that in practice when a buoyant gas is released some way below the ceiling, although the mechanism suggested by Leach and Bloomfield may operate initially, a layer of essentially uniform concentration is formed very quickly between the point of leakage and the ceiling. Whilst the average concentration will continue to increase, the data demonstrates that the general shape of the concentration profile will be maintained throughout the duration of the gas release, as shown in Figure 3.

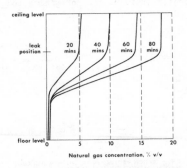

Figure 3 DEVELOPMENT OF A CONCENTRATION PROFILE
WITH TIME (UNVENTILATED ROOM)

Both the depth of the layer formed and the degree of mixing which occurs between gas and air can be affected by factors other than the leak position within an enclosure. These include the gas escape velocity, the volume flowrate of gas and the leak orientation.

In Figure 4 the degree of mixing, characterised by the ratio of the maximum gas concentration in the layer to the average gas concentration in the enclosure, is shown as a function of both leak position and orientation. (As better mixing occurs the value of the ratio decreases: perfect mixing corresponds to a value of unity). It is evident from Figure 4 that better mixing is achieved with downward pointing leaks than with either horizontal or upward orientated leaks. As might be expected, a similar trend is observed in the variation of layer depth with leak orientation, as illustrated in Figure 5.

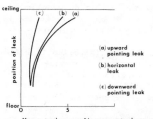

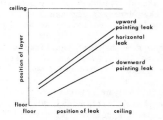

Max. natural gas conc⁰/average natural gas conc⁰

Figure 4   DEGREE OF GAS MIXING CHARACTERISED BY RATIO OF MAXIMUM TO AVERAGE GAS CONCENTRATION AS A FUNCTION OF LEAK POSITION AND ORIENTATION

Figure 5   POSITION OF NATURAL GAS LAYER AS A FUNCTION OF LEAK POSITION AND ORIENTATION

Tests in the smaller, rectangular shaped vessel indicated that the position of the source of leakage within an enclosure can have a slight effect on the concentration profile developed, particularly in the early stages of the mixing process. Thus, for the same leak source characteristics, it was observed that initially there was marginally less mixing with the leak source located against the wall of the vessel than when gas was released centrally in the chamber, i.e. the concentration at the ceiling was higher, with less gas present lower down in the vessel, compared with the centrally located leak. However, this slight difference between the two cases diminished with time and with a leak source located half way down the test chamber ($z = 0.5$) the difference in concentration profiles from the two situations was negligible after a gas input equivalent to approximately 5% of the vessel volume. The location (as against the height) of a leak is not considered therefore to have any real practical significance.

The volume flowrate of gas and the gas escape velocity also affect the degree of mixing between gas and air. Higher gas velocities tend to promote more mixing and lead to deeper layers whereas the volume flow rate of the gas released affects the time taken to reach a given concentration - the higher the flow rate, the less time required to reach a specified concentration.

Experiments in a multi-room test building produced results similar to those obtained in the 20.6m³ test chamber. Thus, it was observed that, if the leak position was below the level of the door lintels, although the room into which gas was released had a slightly higher gas concentration at ceiling level, below the level of the door lintels the gas concentration was similar in each room. In addition, for a leak position below door lintel level, gas initially did not accumulate at ceiling level in the room in which it was released and then pass into adjacent rooms. Only for a leak position higher than the door lintels did the gas build up at ceiling level before spilling under the lintel into an adjacent room. It would appear therefore that the mixing behaviour observed in a single cell enclosure can describe adequately for practical purposes the behaviour of gas released into more complex geometry enclosures.

Ventilated Situation

When ventilation is present, an additional process is available to promote mixing; this is the turbulent interchange between escaping gas and the air supplied by ventilation. In many situations, mixing and subsequent dilution of the gas mixture by the action of ventilation will be the dominant process by which the build up of a flammable gas concentration following an accidental release will be prevented.

Although mechanical ventilation systems are often installed in buildings, in most situations the requisite air change rate is provided by natural ventilation. The driving forces for natural ventilation are (a) the pressure differentials created across a building by external wind forces and (b) buoyancy forces derived from the difference in densities of the atmospheres within and outside a building. Normally, buoyancy driven ventilation derives from a difference in temperature between the air inside a building and that outside but it can be driven by density differences caused by the release of a buoyant gas within a building. The two common patterns of ventilation (i.e. wind driven and buoyancy dominated) are illustrated in Figure 6. Methods for calculating natural ventilation rates of buildings are described in several publications, for example British Standard 5925[2].

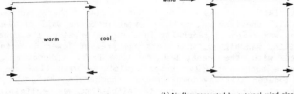

(a) Air flow generated by temperature difference alone
Upward air flow ventilation

(b) Air flow generated by external wind alone
Cross flow ventilation

Figure 6   COMMON PATTERNS OF VENTILATION

The effect of different patterns and rates of ventilation on gas mixing and accumulation have been studied extensively in the 20.6m^3 cubical test vessel. These experiments were carried out under controlled air flow conditions to ensure consistency between tests. However, tests of a similar nature were made also in naturally ventilated buildings to provide a comparison of the pattern of gas mixing and accumulation under more practical ventilation conditions.

Each of the ventilation regimes investigated tended to produce different shaped concentration profiles. This was most obvious between the normal, upward ventilation air flow and the unusual reverse flow situation of a downward air flow. Concentration profiles obtained from the cross flow ventilation regimes tended to approximate to those produced by one or other of the two extremes of upward and downward air flow, depending on the location of the inlet and outlet ventilators. Each of the different air flow regimes is discussed separately below.

i) Upward Ventilation

The experiments with upward ventilation demonstrated that, over a wide range of leak source characteristics and ventilation rates, a steady state concentration profile will be established. The shape of the concentration profile is similar to that obtained under zero ventilation conditions and is maintained as time increases. Typical examples are shown in Figure 7, which indicates that a well defined layer of gas-air mixture, of essentially uniform composition, is formed between the level of the leak and the ceiling with little or no gas being present below the level of the leak.

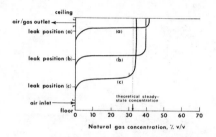

Figure 7   THE EFFECT OF LEAK POSITION ON STEADY-STATE CONCENTRATION PROFILES WITH UPWARD VENTILATION AIR FLOW (NATURAL GAS)

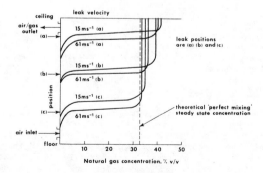

Figure 8   EFFECT OF LEAK VELOCITY AND POSITION (NATURAL GAS) FOR UPWARD FLOW VENTILATION

It was evident from these experiments with ventilation that the nature of the leak source had a similar effect on the degree of mixing as in the unventilated situation. Thus, Figures 7 and 8 demonstrate that better mixing occurs with a lower leak position and higher gas leak velocity.

If perfect mixing between the leaking gas and the ventilation air is
assumed to occur, the build up of gas concentration with time is
described by the equation:

$$C(t) = 100\ Qg/(Qa + Qg)\ [1 - \exp\ (- \frac{(Qa + Qg)t}{V})\ ]\qquad\qquad 1$$

The steady state concentration is given by:

$$C_s = 100\ Qg/(Qa + Qg)\qquad\qquad 2$$

The value of C_s for the experimental results presented in Figures 7 and
8 is indicated by the dotted line. It is evident that the lower the leak
position, the nearer the approach to this 'theoretical' concentration, in
other words, the better the degree of mixing. In practical terms,
however, for leak sources located below the level of the outlet
ventilators, the variation of the steady state concentration from that
predicted by equation (2) is not significant.

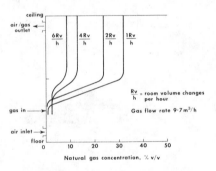

Figure 9   EFFECT OF VENTILATION RATE ON GAS CONCENTRATION
- UPWARD FLOW VENTILATION (NATURAL GAS)

The effect of increasing the rate of ventilation - other factors
remaining constant - is to reduce the maximum concentration and the depth
of the layer of gas-air mixture formed (Figure 9). Overall, the effect
of an upward ventilation air flow on gas mixing and accumulation is to
reinforce the behaviour observed under zero ventilation conditions.

ii) Downward Ventilation

Although in most practical situations downward air flow would be
considered to be an unusual ventilation pattern, it does sometimes occur.
For this ventilation regime, buoyancy and momentum forces are acting
against each other, consequently it is to be expected that better mixing
will result. This is confirmed by the experimental results, Figure 10,
which show that a substantial amount of gas can be present below the
level of the leak in this situation. In many of these experiments, at
steady state conditions, the gas concentration below the level of the
leak position approximately equalled the theoretical value given by
equation (2). However, the shape of the steady state profile is not
established quickly but continually changes as gas accumulates below the
level of the leak. Essentially with a downward air flow regime two
layers are formed: one of a higher concentration above the leak position

and a second with a lower concentration below the level of the leak. This can also be seen in Figure 10.

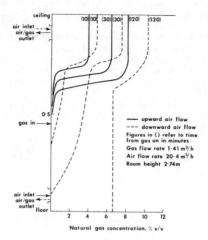

Figure 10   BUILD UP OF GAS CONCENTRATION WITH TIME -
CONCENTRATION PROFILES FOR UPWARD AIR
FLOW AND DOWNWARD AIR FLOW (NATURAL GAS)

iii) Cross Flow Ventilation

A number of cross flow ventilation patterns have been investigated experimentally. Examples of gas concentration profiles at steady state condition for these various ventilation air regimes are presented in Figure 11. Although the shape of the concentration profile depends markedly on the dispositions of the air inlets and outlets, certain trends important to the practical situation can be identified. Thus, the

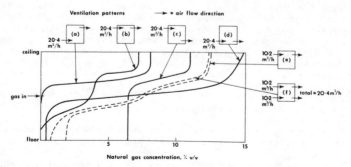

Figure 11   STEADY STATE GAS CONCENTRATION PROFILES FOR DIFFERENT VENTILATION PATTERNS
(Natural gas, flow rate 1·42 m³/h for all cases)

lowest concentrations were always obtained with a high level outlet position, irrespective of the location of the air inlet(s). The data also demonstrates that, as might be expected for a leakage of a buoyant gas, ventilation inlets and outlets at floor level only are almost completely ineffective in dispersing mixture accumulation at high level (pattern (d), Figure 11). In fact, the gas concentrations measured with this configuration were very similar to those obtained under zero ventilation conditions. There was a similar effect with other patterns which included a low level inlet and low level outlet. This was especially noticeable with true cross flow conditions, i.e. inlets and outlets at both high and low level (pattern (f), Figure 11). The concentration profile produced under this ventilation condition was very similar to that obtained with inlets and outlets at high level only (pattern (e), Figure 11) but with only half the ventilation air rate.

The data from these cross flow ventilation experiments imply that for a cross flow situation with inlets and outlets at both high and low level, only the air entering at high level plays a part in the mixing process. This result is important in naturally ventilated buildings and suggests that, whilst low level openings are necessary to provide inlets for buoyancy driven ventilation, calculation of ventilation air openings based on wind driven ventilation should be restricted to high level openings only.

PRACTICAL IMPLICATIONS

The main feature to emerge from the experimental study of gas mixing and accumulation is the tendency of buoyant gas releases to result in the formation of high level layers of gas-air mixture, for a wide range of ventilation conditions. This type of behaviour has several important practical implications for the design of natural ventilation systems of buildings.

The formation of a layer inhibits the mixing of gas with the total volume of air in an enclosure and can, for example, radically influence the time taken for an explosive mixture concentration to be reached following a release of gas. Hence, whilst the data from both the controlled air flow studies and the tests in naturally ventilated buildings indicate that, for many practical situations, the perfect mixing equation can be used to determine the steady state gas concentration, in some situations a modification to this formula is needed to describe the build up of concentration to the steady state more accurately. In particular, the data show that for most common patterns of ventilation, the mixing of gas and ventilation air is restricted mainly to the part of the enclosure above the level of the leak. Consequently, the rate of increase in gas concentration can be higher than that predicted by equation (1). This discrepancy can be corrected adequately by using a reduced volume, V^*, in the perfect mixing equation rather than the actual enclosure volume (V) where V^* is the volume of the enclosure above the level of the leak.

For dense gases and vapours, such as propane and petrol, it may still be more appropriate to use the unmodified perfect mixing equation, the reason being that with these substances the tendency will be to form a layer of mixture at low level rather than a high level ceiling layer as is the case with the release of a buoyant gas. Thus, under normal ventilation conditions, the pattern of gas mixing and accumulation subsequent to a leakage of dense gas should correspond reasonably closely to the behaviour of a buoyant gas under a downward ventilation air flow. In this case, the data indicate that gas-air mixture will be distributed throughout the enclosure volume (pattern (c) Figure 11).

Once formed, a layer of rich gas-air mixture is difficult to disperse, relatively high ventilation rates being required to disturb sufficiently the boundary between the layer and the air in order to allow mixing and, therefore, dilution to occur.   Hence, a primary requisite of any ventilation system must be to prevent the formation of gas rich layers. The experimental data show that for the most effective dispersal of a leakage of buoyant gas, ventilation openings - certainly the outlets - should be located at high level.   Further, data from the cross-flow ventilation studies, with inlets and outlets at both high and low level, indicate that the low level air flow has little or no effect.   This suggests that when calculating the ventilation air flow available from wind-driven ventilation, the value of Qa to be used in equation (1) should be based only on the areas of the high level ventilators.

However, this does not mean that openings should not be placed at low level.   In normal circumstances, the natural ventilation of a building will be by wind-driven ventilation and the air change rate will be variable since it will be influenced by both wind speed and wind direction.   Whilst variations in wind characteristics can be accommodated by distributing openings on all sides of a building, ideally these should also be located both at high and low level.   This will then ensure that a reduction in ventilation rate, due for example to a drop in wind speed, can be minimised by the buoyancy driven ventilation that would then occur.   (Below a wind speed of about 2 m/s, buoyancy driven ventilation will tend to be more dominant than wind driven ventilation in many situations).   The contribution of buoyancy effects to the ventilation can be maximised by ensuring the greatest possible distance between the high and low level ventilation openings.   By this means, buoyancy driven ventilation will be available when required and the ventilation air flow rate will depend less on wind characteristics and be less variable than otherwise.

CONCLUSIONS

The main conclusions to be drawn from the experimental study are:

1. Over the wide range of ventilation conditions investigated there is a tendency for gas releases to result in the formation of a layer of gas-air mixture in part of an enclosure, rather than the dispersal of gas throughout the entire volume.

2. The steady-state gas concentration following an accidental release is described adequately by the perfect mixing equation.   A modified version of this formula should be used to calculate the build-up of gas concentration to steady state conditions.

3. The most effective dispersal of mixture accumulation is obtained with ventilators placed at high level but to minimise variations in the air flow rate available from natural ventilation systems, ventilation openings should be located at both high and low level on all available faces of a building.

NOMENCLATURE

$C(t)$	–	time dependent gas concentration (% gas in mixture)
C_s	–	steady state gas concentration (% gas in mixture)
Q_a	–	air flow rate (m^3/hr)
Q_g	–	gas leakage rate (m^3/hr)
t	–	time (hr)
V	–	enclosure volume (m^3)
$V*$	–	enclosure volume above leak (m^3)
v	–	gas leakage velocity (m/s)
z	–	leak position in enclosure (dimensionless)

The leak position is defined in terms of the height of the enclosure: a value of $z = 1.0$ indicates a leakage at ceiling level, a value of $z = 0$ corresponding to a leakage of floor level.

REFERENCES

1. Leach, S.J. and Bloomfield,D.P., Ventilation in Relation to Toxic and Flammable Gases in Buildings, Building Science, **8**, 289 (1973)

2. British Standard 5925, Code of Practice for Design of Buildings. Ventilation Principles and Designing for Natural Ventilation (formally CP3 Chapter 1(c)). British Standards Institute (1980)

EXPERIMENTAL INVESTIGATIONS ON THE RUN-UP DISTANCE OF GASEOUS DETONATIONS IN LARGE PIPES

BY H. STEEN AND K. SCHAMPEL

PHYSIKALISCH-TECHNISCHE BUNDESANSTALT, BRAUNSCHWEIG

SUMMARY

For the design of installations in which explosive mixtures of flammable gases and vapours with air can occur, the run-up distance in pipes is a significant quantity for safety engineering. For propane/air and ethylene/ air mixtures it was measured by varying a number of influencing parameters (such as, for example, concentration of the flammable component, pipe diameter, type and location of the ignition source, pre-volume, turbulence, and pipe blockage). Though the tests carried out do not yet suffice to make a generalized quantitative statement on the run-up distance, some aspects shown by the results are of interest to the application in industrial practice: thus, the strength of the ignition source is of almost no importance, whereas the laminar burning velocity (and thus the maximum experimental safe gap) and, in addition, a volume preceding the pipe plays a considerable role.

INTRODUCTION

Owing to the high, though only short duration momentum, the occurrence of detonations of explosive vapour/air or gas/air mixtures in pipes represents a serious threat to the industrial installations concerned, the more so as they can exceed the stability of the latter /1/. In such cases, it is most interesting from the safety engineering point of view to know under what design conditions (for example, the pipe dimensions) a deflagration can develop into a detonation and, in particular, what run-up distance (distance between location of ignition source and location of deflagration/detonation transition) is necessary for a detonation to take place. Knowledge of the run-up distance is therefore a prerequisite for taking, in particular, constructive measures against the occurrence and effects of detonations and the ensuing high dynamic loads acting on the installations concerned. These protective measures comprise the use of only small pipe lengths or, on the other hand, the use of flame-arresting devices which stop the flame propagation in the case of a detonation (detonation-arresting devices).

To the question in what pipe lengths detonations must be reckoned with, the applicable safety regulations (at least in the Federal Republic of Germany) give no reply or only a very general one. In the "Explosions-schutz-Richtlinien"(Explosion Protection Guidelines) /2/, a pipe length L to diameter D ratio of at least L/D = 5 is given; owing to its necessarily general nature, this limiting value should, however, be regarded as conservative. Moreover, the findings hitherto published with respect to the run-up distance of detonations, particularly in the case of large pipe diameters and under different conditions of installation in industrial plants (for example, the provision of a large volume), are

still incomplete. The experimental investigations carried out in the
Physikalisch-Technische Bundesanstalt are thus aimed at providing the
designers of industrial plants with more detailed aids when taking protec
tive measures against detonations.

When very high ignition energies are applied a detonation can be initiat
without a pronounced and prolonged initial phase, the ignition energies
required being of the order of at least some Joule in the case of a spark
ignition of acetylene/oxygen mixtures /3/ and of the order of 50 g of explosive in the case of chemical ignition of methane/air mixtures /4/.
What will, however, be considered and investigated here is not such a
direct initiation of a detonation (with a run-up distance of only a few
centimetres) but the run-up process with a "weak" ignition source (i. e
with an ignition source which does not produce any substantial shock wave
/5/. For mixtures of acetylene and ethylene with oxygen and nitrogen (but
not with air), Wagner et al. /6/ proved that the flame is very rapidly
accelerated and develops into a detonation where there are grids and othe
obstacles: likewise producing turbulence. Tests by other authors furnish
similar information. As will be shown below, such a direct initiation of
detonations will not ensue within the scope of the tests carried out here
with propane/air and ethylene/air mixtures and without such obstacles in
the pipe even if there is a pre-volume.

KNOWLEDGE ACQUIRED TO DATE

A theoretical model of the run-up process of detonations which would be
suitable as a basis for safety engineering considerations and could be
quantitatively evaluated with sufficient accuracy is not yet available.
As a whole, the test results published with respect to the run-up distan
cannot be classified systematically, as the respective test conditions
(for example, the kind of detonative mixtures, pipe geometry, pressure a
temperature) differ widely and were probably selected with a view to
answering specific questions.

It is a prominent feature of most of the tests published that they were
carried out on relatively small pipe diameters /7 to 12,15/. Bollinger
et al. /7 to 9/ tested pipes 9 mm, 15 mm, 50 mm and 74 mm in diameter,
using mixtures of hydrogen, methane, carbon monoxide with oxygen and of
acetylene with oxygen and nitrogen at temperatures of 40 $^{\circ}$C to 200 $^{\circ}$C; i
addition, they investigated in particular the influence which the mixture's temperature and pressure exert on the run-up distance. After tests
with nitrogen/oxygen at atmospheric pressure and temperature values in
pipes up to 51 mm in diameter, Oppenheim et al. /10/ obtained the relationship between the run-up distance s_D and the pipe diameter D;
$s_D = 47 \cdot \sqrt{D}$.

On the other hand, in their tests carried out with ethane/oxygen in a
pipe with an internal diameter of 10 mm, with a view to determining the
influence of the flow conditions of the unburnt mixture on the run-up
distance, Baumann ånd Wagner /11/ found the realtionship $s_D \sim Re^{-0.46}$.
This needs not necessarily to be contradictory to the result Oppenheim
et al. found, as their tests were carried out with a mixture at rest and
Baumann's and Wagner's statement is valid only for a definite pipe diameter. In this connection, some previous measurements carried out by
Laffitte /12/ on pipes up to 51 mm in diameter with mixtures of carbon

disulphide vapours and oxygen, are of interest: they show a linear dependence of the run-up distance s_D upon the diameter D - but only for diameters of more than some 25 mm.

Hattwig /13/ recently published measurements of the run-up distance for various mixtures of hydrogen, methane, nitrogen and nitrogen oxide in nominal pipe widths of 50 to 500 mm; his measuring results, too, show a linear relationship between the run-up distance and the pipe diameter. Such a linear relationship results also from the measurements by Bartknecht /14/ with methane/air and hydrogen/air mixtures in pipes up to 400 mm in diameter. This linear relationship thus indicates that the pipe line's run-up distance/diameter ratio s_D/D assumes a constant value for major pipe diameters (D > 25 mm). This value is, however, certainly dependent on a series of other influencing parameters. The ratio s_D/D

(a) is lowest with fuel/air or fuel/oxygen mixtures which show the maximum laminar burning velocity (i.e. more or less in the neighbourhood of the stoichiometric mixture)

(b) decreases with increasing pressure /7,8,15/

(c) increases with increasing temperature of the unburnt mixture /7/

(d) decreases with increasing turbulence (and thus increasing Reynolds number) of the flow in the pipe /9,11/.

As, in this connection, only the atmospheric conditions which are of interest to most of the practical cases are considered, the experimental findings, especially those given under (a) and (d), are of particular importance and will be dealt with in the following in more detail:

Among other things, on the basis of their measurements of the run-up distance, Ginsburg and Bulkley /15/ pointed out the influence of the laminar burning velocity Λ : The larger Λ, the smaller the run-up distance s_D; the scatter of their measurement values is, however, so pronounced that a clear dependence cannot be derived. Bollinger et al. /7/, too, indicate such a dependence. The clarity of the parameters influencing the run-up process and thus the development of a theoretical model that can be evaluated quantitatively are still impeded by the influence of the flow process /9,11/. That is why the tests described here are made with mixtures at rest and in pipes which are technically smooth. Pawel, Schampel and Schön /16/ tested methane/air mixtures in large pipes (with nominal diameter of 100 mm and 200 mm) with respect to the maximum attainable flame velocity. Although it was not possible to observe stable detonations, the flame velocity was found comparable to that which is typical for the transition of a deflagration to a detonation in the case of saturated hydrocarbons.

After this general survey it can be stated that as a result of the lack of calculation procedures, systematic experiments which are necessary to acquire knowledge of the run-up distances of detonations have not yet been reported on comprehensively in the literature and are therefore urgently required.

EXPERIMENTAL SET-UP AND METHOD

At the closed end of technically smooth pipes of various diameters and of sufficient length for the run-up process, an ignition was initiated and the flame propagated towards the opposite open pipe end which was closed with a thin plastic foil during rinsing with the test mixture. For some tests, the closed pipe end was preceded by an additional pre-volume of varying size in which the ignition was initiated. The passage of the flame (recorded with a transient recorder) was registered by ionization probes, each spaced at a distance of 2 m. The records furnished flame propagation velocities averaged over the 2 m distance. The point of intersection of the curve of non-stationary flame velocity (deflagration) in the initial stage with the stationary detonation velocity is referred to as the point of transition of deflagration to detonation and, as a result, the distance between the flange at the beginning of the pipe and this point is denoted as run-up distance s_D of the detonation (see Fig. 1).

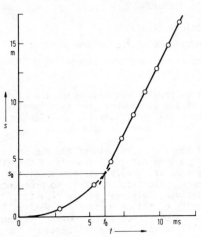

The tests were carried out with explosive mixtures of ethylene (99,7 % purity) and technical propane (appr. 5 % impurities, mainly ethane and butane) with air. The mixtures were produced in the continuous flow process. The concentration of the fuel component was controlled on account of the mixtures maximum experimental safe gap (MESG) (whose dependence on fuel concentration was previously determinded in the laboratory with the aid of a standardized MESG measuring device /17/) and additionally with an infrared gas analyzer.

In the tests the following sources of ignition were used (for exceptions, see Section 4):

Fig. 1:
Distance/time diagram with transition point deflagration/detonation of ethylene/air mixture (6,5 % by volume), pipe ND50

Type A: fusing wire igniters, electrical resistance 1.4 to 1.8 Ω , operated at 35 V (a.c.)

Type B: chemical igniters (10 kJ) producing no substantial shock wave but only an ignition flame which is larger than that of type A.

TEST RESULTS AND THEIR DISCUSSION

The test results obtained for the run-up distance of the detonation are mean values from several measurements (except for the results obtained for the large pipe diameters of 250 and 400 mm); the relative maximum deviations from these mean values were of the order of some 15 %. Apart from a certain statistical spread, the measuring uncertainty is due to relatively large spacing of the probes and the relatively small difference of slopes of the distance/time diagram for the deflagration and for the detonation at the moment of transition (see. Fig. 1).

In accordance with the statements of other authors, the minimum of the run-up distance s_D resulted for the composition of the most easily ignitable mixture; this can be seen from Fig. 2 in which the run-up distance s_D for propane/air and ethylene/air mixtures is shown in dependence upon the ratio of the concentration to the stoichiometric concentration for a pipe of 50 mmm nominal diameter without pre-volume. The fact that a pronounced minimum as with propane is not obtained with ethylene is probably due to the decomposition reaction of ethylene at high concentrations. Further tests carried out with a view to clarifying the various influencing parameters were performed with the most easily ignitable composition.

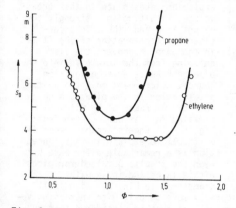

Fig. 2:
Run-up distances s_D depending on equivalence ratio $\emptyset$ for ethylene and propane/air mixtures. Ignition at closed end of pipe ND 50 with $V_B \sim 0$ m^3.

Fig. 3 shows the dependence of s_D upon the diameter of the pipe. Its linear realtionship confirms the qualitative model considerations by Shchelkin and Troshin /18/ and the measurement results obtained by Hattwig /13/ and also by Bartknecht /14/.

During the tests described, the mixture was at rest prior to ignition. From many experimental investigations, however, the strong influence of turbulence upon flame acceleration is known, which thus would have to produce a decrease of the run-up distance s_D. For this reason, the influence exerted by the turbulence of the flowing mixture was also investigated (see Fig. 4). It must, however, be stated that in our tests, the turbulence of the flowing mixture was relatively weak, as it was caused only by the flow in the pipe with the usual technical roughnesses. This certainly represents a restriction to the evidence provided by the test

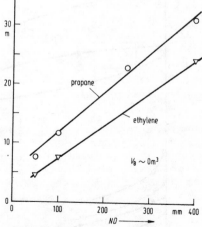

Fig. 3: Run-up distance s_D for most easily ignitable ethylene and propane/air mixture depending on pipe diameter ND and pre-volume $V_B \sim 0$ m^3.

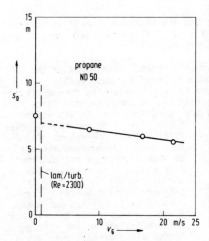

Fig. 4:
Run-up distance s_D of propane/air mixture in a pipe ND 50 depending on flow velocity v_G of unburnt mixture

results. In any case, these results show only a very small dependence upon the flow velocity which at most yielded only a Reynolds number of some 75 000.

The influence of the pre-volume V_B (see Fig. 5) preceding the pipe shows the same tendency as that of the turbulence: After ignition close to the wall, a relatively slow deflagrative explosion takes place in this pre-volume, resulting in a strongly turbulent flow and, thus, in a considerable acceleration of the flame in the pipe. Moreover, the flame entering from the volume V_B into the pipe represents a large ignition source, which means that the initial acceleration in the pipe is high. Thus the decrease of s_D with increasing V_B - as can be seen from Fig. 5 - is understandable. What is difficult to explain in this connection is a remarkable difference between the results obtained for pipes with a large and those with a small diameter (see Fig. 5): with small pipe diameters, s_D decreases considerably more quickly with increasing V_B than in the case of large pipes. However, it must be considered that on

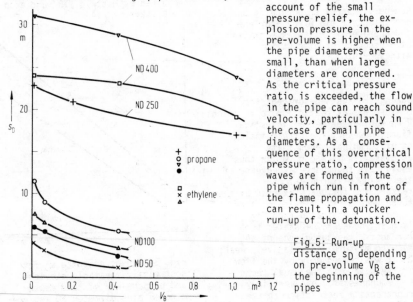

account of the small pressure relief, the explosion pressure in the pre-volume is higher when the pipe diameters are small, than when large diameters are concerned. As the critical pressure ratio is exceeded, the flow in the pipe can reach sound velocity, particularly in the case of small pipe diameters. As a consequence of this overcritical pressure ratio, compression waves are formed in the pipe which run in front of the flame propagation and can result in a quicker run-up of the detonation.

Fig.5: Run-up distance s_D depending on pre-volume V_B at the beginning of the pipes

In practice, in particular the experimental finding is of importance that the run-up distance does not decrease steadily with increasing pre-volume V_B but that it approaches a constant final value. This observation is in line with the above interpretation, as the compression waves are restricted by the maximum explosion pressure in the pre-volume V_B.

Within the scope of these measurements - in particular on the pipe diameters of 50 mm and 100 mm -, some tests were carried out with a detonation arrester mounted at the end of the pipe. The influence of these arresters on the run-up distance was small, though not insignificant. For this reason, the values s_D in Fig. 2 (for the most easily ignitable mixture) and those given in Fig. 5 (for $V_B \sim 0$ m³) some of which were obtained with a flame arrester at the pipe end are not entirely in agreement.

The initial conditions of flame acceleration in the deflagrative run-up phase are certainly also determined by the kind of the ignition source used. In some of the tests carried out to clarify the dependence of the run-up distance upon the concentration of the fuel component according to Fig. 2, it was shown that when using a chemical igniter with a relatively high nominal energy (though without an important shock wave, igniter type B) in the pipe (without pre-volume V_B), the flame accelerated very strongly in the initial phase (up to some 1 to 2 m) up to the detonation velocity but then decelerated to a slower deflagrative flame propagation; a stable detonation, however, still did not occur. This statement is confirmed by the more systematic test results in Fig. 6: In a pipe with a nominal diameter of 150 mm where the ignition took place in a pre-volume $V_B = 0,12$ m³ and which was provided at its end with a detonation arrester D, a considerable intermediate acceleration resulted in the initial phase when the strongest igniter (curve 4) was used. However, the run-up distance s_D was the same for all igniters.

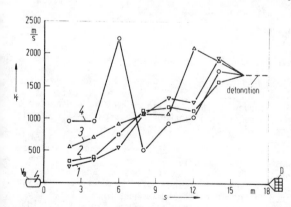

Fig. 6:
Flame velocity v_F during detonation initiation for most easily ignitable propane/air mixture; pipe ND 150; $V_B = 0,12$ m³

D	detonation arresting device	
1	fusing wire igniter;	500 Ws
2	" " " ;	5000 Ws
3	capsuled explosive igniter;	3000 Ws
4	strongly capsuled explosive igniter;	3000 Ws

Likewise, the influence of the location of the ignition source on the run-up distance was tested experimentally. The test showed that in the case of both propane/air and ethylene/air mixtures, a high flame acceleration up to detonation does not result when igniting at the open end of the pipe (see /19/). An interesting aspect of the ignition source location was shown by tests in which the ignition source was not located at the beginning or at the end but somewhere in the pipe. In a ND 50 pipe of varying length whose ends were both open (closed only with a plastic foil, see Fig. 7), ignition was initiated in various points so that in all run-up tests the pipe was in one direction of equal length as seen from the ignition source, whereas it varied between 6 m and 18 m on the other side. A detonation was produced only when on both sides at least two of the 6 m long pipe sections were connected.

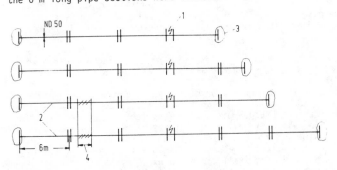

Even when the open pipe end was partially blocked - the ignition source being situated directly in front of the orifice within the pipe - substantial flame accelerations could not be observed during tests on a pipe with a nominal diameter of 50 mm and with orifices with a free diameter of

Fig. 7:
Experimental set-up for ignition inside pipes of ND 50

1 location of ignition source
2 pipe length; 6 m each
3 open end covered with plastic foil
4 measured location of transition from deflagration
   to detonation

13,5 mm, 15 mm and 22 mm. However, with an orifice with a free diameter of only 12,5 mm - i.e. with a pipe section blockage of approx. 94 % - detonations occurred almost unexpectedly, but reproducibly.

It might be of great interest to practice to know how the run-up distance depends upon the properties of the explosive atmosphere, i. e. upon the kind of the fuel component. This raises also the question whether the experimental results obtained here and the conclusions drawn from them also apply to mixtures of other flammable gases and vapours with air. On the basis of more or less qualitative considerations, Shchelkin and Troshin /18/ come to the conclusion that the run-up distance s_D is proportional to $1/\Lambda^2$ (Λ being the laminar burning velocity). According to Bollinger et al., s_D is proportional "to the burning velocity raised to some empirically determined negative power" /8/. The tests with propane and ethylene carried out here certainly do not yet suffice to give an undisputed description of the substance dependence, but the quadratical dependence described by Shchelkin and Troshin is not confirmed by tests, which is indicated by Fig. 2. According to this the dependence of the run-up distance s_D from the laminar burning velocity to be assumed is susbstantially weaker. What is indisputable is the statement that a mixture with a low laminar burning velocity results in a longer run-up distance s_D. For the practical application of this statement the laminar burning velocities for a sufficient number of gas/air or vapour/air mixtures are, however, not available. Instead of this, it is possible to use the maximum experimental safe gap (MESG) which has already been determined for a rather large number of mixtures and is relatively easy to determine on account of the standardized measurement method /17, 20/. For a large number of flammable gases and vapours a relationship (after data treatment by method of least squares) between and the maximum experimental safe gap was established, which can be described sufficiently well by $\Lambda \sim 1/MESG$.

In some tests not only the run-up distance s_D but also the run-up time t_D was measured. It was shown that - all other conditions being unchanged - the ratio s_D/t_D which is also referred to as Oppenheim velocity was almost (i.e. to a measuring uncertainty of some $\pm$ 20 %) independent of the concentration of the fuel component and that it was almost equal for the mixtures with propane and ethylene; in the measurement results given in Fig. 2 for a pipe diameter of 50 mm it was some 580 $\pm$ 100 m/s. In the tests with larger pipe diameters, it was noticeably smaller (for ND 250, for example, by the factor 5). This statement is inconsistent with the observations of Oppenheim et wa. /10/ and Baumann and Wagner /11/, who established a practically constant Oppenheim velocity for pipe diameters of less than 5 cm. Obviously the self-turbulence of the flame which is smaller in the case of large pipe diameters in particular in the initial phase of propagation, effects a slower flame acceleration which became noticeable mainly in the considerable extension of the run-up time.

SOME PRACTICAL CONCLUSIONS PERTAINING TO SAFETY ENGINEERING

The run-up distance s_D of detonations is subject to the influence of several parameters (such as pre-volume V_B, turbulence, changes in cross section) so that the values measured for straight, technically smooth pipe lines without obstacles cannot be applied to practical conditions without a considerable safety factor. A value of this factor which would be applicable to all possible conditions is almost impossible to establish.

On the basis of the measurement values obtained here for propane and ethylene, it is not yet possible to make an undisputed statement as to the run-up distance of mixtures of other flammable gases and vapours. To achieve this, further experiments must be carried out. It can, however, be stated, that a mixture with a lower laminar burning velocity or a larger maximum experimental safe gap MESG yields a longer run-up distance s_D.

The run-up distance is hardly ever influenced by the strength of the ignition source but is considerably reduced with increasing size of the volume preceding the pipe. The latter is of particular importance in the design of industrial plants, as in many cases, the inside of various installations such as some chemical plants is interconnected by pipes; in these cases, when the flame is propagated after ignition in one piece of installation, the chance of the run-up of a detonation in the connecting pipe is dependent, among other things, on its length.

R e f e r e n c e s

/1/ H. Steen, K. Schampel: Zur Druckbeanspruchung detonationssicherer
Einrichtungen, PTB-Mitteilungen 92, p. 32-38 (1982)

/2/ "Explosionsschutz-Richtlinien (Ex-RL)" (Guidelines für Explosion
Protection) edited by Berufsgenossenschaft der Chemischen Industrie,
Gaisbergstr. 11, D 6900 Heidelberg, 1980

/3/ R. Knystautas, J.H. Lee: On the Effective Energy for Direct
Initiation of Gaseous Detonation, Combustion and Flame 27 p. 221-228
(1976)

/4/ S.M. Kogarko: Detonation of Methane-Air Mixtures and the Detonation
Limits of Hydrocarbon-Air Mixtures in a Large-Diameter Pipe,
Soviet Physics-Technical Physics 3 p, 1904 ff (1958)

/5/ W. Jost, H.Gg. Wagner in: Handbuch der Raumexplosionen (Editor:
H.H. Freytag), Section 1b Verlag Chemie GmbH, 1965

/6/ K.J. Dörge, D. Pangritz, H.Gg. Wagner: Experiments on velocity
augmentation of spherical flames by grids, Acta Astronautica 3 ,
p. 1067 (1976)

/7/ L.E. Bollinger, M.C. Fong, R. Edse: Experimental Measurements and
Theoretical Analysis of Detonations Induction Distances,
ARS-Journal 31 , p. 588 - 595 (1961)

/8/ L.E. Bollinger: Experimental Detonation Velocities and Induction
Distances in Hydrogen Air Mixture, AIAA-Journal 2 , p. 131-133 (1964)

/9/ L.E. Bollinger, G.C. S.mith, F.J. Tomazic, R. Edse:
Formation of Detonation Waves in Flowing Hydrogen Oxygen and
Methane-Oxygen Mixtures, AIAA-Journal 4, p. 1773-1776 (1966)

/10/ W. Baumann, PA. Urtiew, A.K. Oppenheim: On the Influence of Tube
Diameter on the Development of Gaseous Detonation
Zeitschrift für Elektrochemie 65, p. 898-902 (1961)

/11/ W. Baumann, H. Gg. Wagner: Einfluß der Frischgasströmung auf die Be-
schleunigungen von Flammen und den Anlaufvorgang von Detonationen
Zeitschrift für Elektrochemie 65, p. 895 - 898 (1961)

/12/ P. Laffitte: Sur la formation de l'onde explosive, Comptes Rendus
(hebd., Paris) 176, p. 1392 - 1395 (1923)

/13/ M. Hattwig: Detonationsanlaufstrecken von Gasgemischen in Rohren
großen Durchmessers, Amts- und Mitteilungsblatt der Bundesanstalt
für Materialprüfung (BAM), Berlin, 10, p. 274 - 278 (1980)

/14/ W. Bartknecht: Brenngas- und Staubexplosionen
Forschungsbericht F 49 (1971) des Bundesinstituts für Arbeitsschutz
(now: Bundesanstalt für Unfallforschung und Arbeitsschutz, Dortmund)

/15/ I. Ginsburgh, W.L. Bulkly: Hydrocarbon-air detonations ...
     industrial aspects,
     Chem. Engng. Progr., Vol 59, p. 82 . 86 (1963)

/16/ D. Pawel, K. Schampel, G. Schön: Gefahren durch erhöhte Flammenge-
     schwindigkeiten, Übergang Explosion zu Detonation
     Physikalisch-Technische Bundesanstalt, Jahresbericht 1976, S. 140

/17/ T. Redeker: Classification of Flammable Gases and Vapours by the
     Flameproof Safe Gap and the Incendivity of Electrical Sparks,
     Physikalisch-Technische Bundesanstalt, Report W-18 (1981)

/18/ K.I. Shchelkin, Y.K. Troshin: Gasdynamics of Combustion,
     Mono Book Corp. Baltimore, 1965

/19/ K. Schampel: Verhinderung eines Dauerbrandes an Flammendurchschlag-
     sicherungen in Lüftungsleitungen von Behältern und Apparaturen,
     2. Sicherheitstechnische Vortragsveranstaltung über Fragen des
     Explosionsschutzes, Physikalisch-Technische Bundesanstalt
     Braunschweig, 10. Februar 1983

/20/ G. A. Lunn: The Influence of Chemical Structure and Combustion
     Reactions on the Maximum Experimental Safe Gap of Industrial Gases
     and Liquids, Journ. Hazardous Mat. 6 (1982), p. 341 - 359)

PRESSURES GENERATED BY BLAST WAVES IN CHANNELS WITH 90° BENDS

P. Fearnley* and M.A. Nettleton[†]
* BP Research Centre, Sunbury-on-Thames, Middlesex
† CERL, Kelvin Avenue, Leatherhead, Surrey KT22 7SE

SUMMARY

Single-shot Schlieren photography has been used to freeze the shapes of initially planar shocks in either air or argon as they propagated around bends in a channel 22 × 47.4 mm. Pressure recordings downstream of the bend have been made, in order to follow the shock re-establishing planarity. By using shocks of different initial strength $M_s \sim 3$, 2 and 1.2 and different radii of curvature of the bend 75, 140 and 300 mm, the influence of shock strength and radius of curvature on the shape of the resultant shock has been determined.

In sharp bends enhancement of shock velocity at the outer wall and its attenuation at the inner wall lead to highly asymmetric stressing of the bend. However, with bends of larger radii of curvature ($R \gtrsim 10r$), the shock tends to propagate around the bend at a constant angular velocity, resulting in a much more uniform distribution of internal stresses. Evidently considerable caution is required in applying the concept of a safe containment pressure to a bend which is potentially subject to the effects of an internal explosion.

For all shock velocities tested and for all bends the overall effect of the bend was to reduce the shock velocity at ≤ 15 diameters downstream of the bend. This is apparently a result of the non-symmetrical interactions of the wave systems produced at inner and outer walls.

List of Symbols

M = Mach number of the shock

R = Radius of curvature of the bend

r = Half-width of the bend

Z = Pressure ratio across the shock (shock strength)

α = Angle between the head of the expansion wave and a projection of the inner wall

θ = Radial angle travelled by the shock in the bend

γ = Specific heat ratio of the test gas

Subscripts

s shock at entrance to the bend

c shock at the outer wall of the bend

e shock at the inner wall of the bend.

1. INTRODUCTION

Industrial pipelines conveying flammable mixtures or gases such as acetylene, which decompose exothermically in the absence of oxygen, usually incorporate changes in cross-section, junctions and bends. In the event of an explosion in such a pipeline, it is evidently important to be able to predict the nature of the stresses occurring both in these features and downstream of them. Whilst there is some information available on the effects of changes in cross-section and of junctions on the strength of the leading shock[1,2,3], much less is known about their influence on the compressed flow behind the front.

This situation is a result of the availability of adequate theoretical treatments of the diffraction of the shock itself (Whitham[4]), involving the Chester-Chisnell[5] relationship between the strength of the shock and its area. The latter incorporates the assumption that disturbances induced in the flow effectively neutralise one another at the front itself. Although much experimental work has been done to confirm this assumption[1,2,3], there has been little work on the effects of the disturbances on the flow behind the front.

Available information on explosions in bends includes a study of shocks propagating through bends with R >> r (where r = radius of the bore and R = radius of curvature[6]) which showed that for one particular velocity of shock at the inlet to the bend, M_s = 2.53, the front remained planar throughout the bend and travelled through it at a constant angular velocity. Milton and Archer[7] have examined the focussing of shocks travelling through logarithmic spirals. Both Wu and Ostrowski[8] and Takayama[9] have studied shocks propagating through 90° bends. More recently Edwards and co-workers[10,11] have given some details of the way in which detonations in mixtures of acetylene-oxygen are influenced by the curvature of a bend. Again it should be emphasised that these studies were mainly concerned with the processes occurring within the bend, so that any information on the resultant front downstream of the bend was a fortuitous outcome.

At this stage it may be helpful to describe the processes involved in the diffraction of shock waves on convex and concave surfaces in the absence of further curved walls, in order to assess the nature of the interactions of the induced waves both in the bend and downstream of it. At the start of the concave surface (the outer bend) the incident shock, velocity M_s, will suffer a Mach reflection, whereby the Mach stem, velocity $M_c > M_s$, propagates orthogonally to the wall and is joined at the triple point to the remainder of the original wave by a reflected shock. The gas close to the wall is compressed by the strengthening Mach stem and delineated by a slip stream from gas compressed by both original and reflected waves. Further along the surface the Mach reflection transforms into a regular reflection, where the reflected wave remains attached to the foot of the original wave in its travel around the bend. At the convex surface (inner bend) a non-centred expansion fan results in the continuous but gradual attenuation of the original front. Thus, a shock profile is formed, with a straight portion of the original front joined to a curved portion starting at the head of the expansion fan and ending with a decaying wall shock travelling orthogonally to the wall.

The interactions of the compressive and expansive wave systems and the resultant systems with the walls, resulting in a stable shock within or downstream of a bend, are evidently complex. Following the interaction of the reflected shock wave with the flow-field associated with the expansion, the resultant waves will reflect from the walls to produce further interactions. Since both enhancement and attenuation of the original front depend on the radius of curvature of the bend and the positions at which the various interactions occur depend on the radius of the channel, there is evidently scope for modifying the stresses and their distribution along the bend by a judicious choice of these parameters.

The present paper shows typical shock profiles in bends of three different radii of curvature in a channel of rectangular cross-section. These have been used to derive the velocities and consequent pressures along the curved channel walls. In addition the pressures associated with the shocked flow have been directly measured at two positions downstream of the bend.

2. EXPERIMENTAL

Planar shocks were generated in either air or argon at sub-atmospheric initial pressures in a 3 m length of 47.4 mm by 22 mm rectangular wave guide tube using air or helium as the high pressure gas. The wave guide channel was connected to the window section (Fig. 1), consisting of two 620 mm diameter Schlieren-quality glass windows, between which two aluminium blocks were sandwiched. These were constructed to allow for the fitting of further blocks to produce bends with nominal values of R of 70, 150 and 300 mm and r = 23.7 mm. Although the inner blocks forming the bend were machined to feather-edges there were slight changes in cross-section of the channel, where these edges met the outer block. These produced very weak reflected compression waves in the flow behind the shock.

A conventional Schlieren system, with 300 mm diameter mirrors to cover as much of the bend as possible, was used. The light source was a Lunatron argon jet operated in a single-shot mode. Consequently, by firing the source after known and increasing intervals, a composite picture of the propagation of the shock through the bend was built from a series of experiments with small but unavoidable variations in the strength of incident shock.

Conventional shock-tube instrumentation was used. Kistler 601 A and B pressure transducers in conjunction with Kistler amplifiers, previously calibrated on shocks of known strength in various pressures of argon, were used to measure shock velocities and any retardation in the straight channel. Pressure records at the numbered stations downstream and upstream of the bend were recorded on 566 Tektronix oscilloscopes. Stations 5 and 6 (15.7 hydraulic diameters from the end of the bend) were on opposite sides of the channel to indicate any non-uniformity in the narrower shock dimension. The light source was fired after a predetermined delay from the incident shock passing station 4 (13.7 diameters from the start of the bend).

3. RESULTS

3.1 Shock Profiles in Bend 1, r = 23.7 and R = 75 mm

Figs. 2, 3 and 4 are composite records of the shock profiles obtained with shocks in air, $M_s \sim 3$, 1.5 and 1.2 respectively, propagating through the sharpest bend. Similar profiles were produced by an $M_s \sim 3$ shock in argon. The dashed lines from the start of the bend represent the paths of the triple point and of the head of the expansion fan, with the straight portion of the profile between the lines being the unmodified portion of the original front. Within the limits of experimental accuracy the angle between the head of the expansion and the wall, α, is within the range predicted from Skew's[12] analysis for shocks of these velocities, $26° < \alpha < 29°$.

With the strongest shock the length of the Mach stem gradually increases in the early stages of the bend (profiles A to E). Thereafter the reflected shock is rapidly convected across the channel, as it encounters the flow-field behind the expansion wave. The profiles downstream of the bend (J, K and L) are highly asymmetric with respect to the cross-section of the channel and will evidently induce a highly complex distribution of stresses. This situation will apparently persist until the triple point reflects from the inner wall.

3.2 Shock Profiles in Bend 2, r = 23.7 and R = 150 mm

Fig. 5 and 6 show composite records of shock profiles for shocks in air $M_s \sim 3$ and $M_s \sim 1.2$ respectively. Similar profiles to those produced by a $M_s \sim 3$ shock were obtained with shocks $M_s \sim 2.1$.

The dimensions of bend 2 approach more closely the Hide and Millar[6] stipulation $R \gg r$ for a shock to propagate with constant angular velocity around the bend, than do the dimensions of the sharpest bend. As their analysis indicates, shocks of $M_s \geq 2.1$ tend to propagate around bend 2 with constant angular velocity. The trajectory of the triple point is a smooth curve and, following its reflection from the inner wall (profile I and J), an approximately planar front is produced within the bend itself. The dimensions of the bend have less effect on the shapes produced from weak shocks. Thus, the profiles obtained from $M_s = 1.2$ shocks indicate that the internal stresses are still highly asymmetric close to the end of the bend.

3.3 Shock Profiles in Bend 3, r = 23.7 and R = 300 mm

Fig. 7 shows the profiles produced by an $M_s \sim 1.2$ shock travelling through bend 3. With incident shocks of higher velocity ($M_s \sim 2.2$ and 3) the resultant profiles approached planarity more closely. Thus, the ratio of velocity at the outer wall to that at the inner wall approaches $\frac{R + r}{R - r}$ as the shock propagates at approximately constant angular velocity around bends with $R/r \geq 12$.

3.4 Pressures Developed Within and Downstream of the Bend

From the preceding arguments, the pressures developed at the walls of sufficiently shallow bends for the shock to travel around them with constant angular velocity, can be calculated from R, r and the standard

Table 1:   Shock Velocities Within and Downstream of Bends

Bend	M_s	M_c	M_e	$M_{5/6}$	$M_{7/8}$
1	2.75 ± 0.02	3.0 ± 0.2 (θ = 19°)	1.94 ± 0.11 (θ = 38°); 1.80 ± 0.±1 (θ = 53°)	2.30 ± 0.05	2.30 ± 0.05
	1.50 ± 0.02			1.28 ± 0.02	
	1.20 ± 0.02			1.17 ± 0.02	
2	3.00 ± 0.03	3.6 ± 0.4 (θ = 11°)	2.55 ± 0.26 (θ = 27°); 1.94 ± 0.14 (θ = 35°)	2.22 ± 0.04	2.17 ± 0.06
	2.10 ± 0.02	2.4 ± 0.2 (θ = 11°)	1.85 ± 0.14 (θ = 27°); 1.83 ± 0.14 (θ = 31°)	2.01 ± 0.01	
3	3.10 ± 0.03			2.70 ± 0.07	2.71 ± 0.07
	2.15 ± 0.02			2.00 ± 0.03	1.58 ± 0.01

relationship between the Mach number and strength of a shock, Z.

$$Z = \frac{2\gamma}{\gamma + 1} M^2 - \frac{\gamma - 1}{\gamma + 1} \qquad \ldots (1)$$

or with M >> 1, the initial internal loads are given approximately by the square of the shock velocity.

Some experimental results for shock velocities at concave and convex walls, M_c and M_e respectively, at distances around bends 1 and 2 represented by the radial angle, θ, are given in Table 1. The velocities were derived from measured distances between shock profiles a known interval apart. The results for the outer wall are confined to values of θ, such that the expansion from the inner wall has not fully interacted with the outer wall shock, whilst maintaining errors in the measurements to ≤10%. The results for the convex wall extend to larger values of θ and suggest that the rate of attenuation of the shock with distance decreases at larger values of θ. The maximum value of the ratio of shock pressure at the outer to that on an inner wall is about 3.

Typical pressure histories at the wider walls of the channel and downstream of the bend are shown in Fig. 8. Stations 7 and 8 are sufficiently close to the closed end of the extension of the channel for the wave reflected from there to return in approximately 0.1 ms. There was essentially no difference in the pressures recorded at opposing stations and all the records indicated the presence of an expansion wave immediately behind the incident shock. Mach numbers derived from such pressure histories are included in Table 1. Incident shocks $M_s > 2.7$ suffered pronounced attenuation in their passage through the bend but stabilised at a constant velocity at or upstream of station 5, at most 15 diameters from the exit of the bend. Shocks of lower Mach numbers continued to decay to produce steep-fronted compression waves in the extension of the bend.

4. DISCUSSION

A study of detonation waves propagating around similarly dimensioned bends[10,11] has shown that the detonation recovers to the Chapman-Jouget velocity at, or prior to the exit of the level, under the influence of the transverse shock waves normally associated with a detonation. Thus, the present results apply to the internal stresses produced by explosions for which the driving force lags behind the leading shock. Typical examples would be a failing detonation front or a classical deflagration with flame and shock widely separated, a localised liquid to vapour explosion and an electrical discharge.

A detailed description of the methods which can be used to predict the pressures at the inner and outer walls and a comparison of predictions and experimental data for both flat and curved walls is given elsewhere[13]. The present paper is intended to call particular attention to the influence of the dimensions of a bend on the uniformity of the load across the bend itself and to the internal loads downstream of the bend.

The present work on blast-waves indicate that, as might be expected, the maximum internal pressure is generated at the outer wall. This contrasts with previous results for detonations. In the latter case, the greatest pressure is generally associated with the re-establishment process, which, depending on the spacing of the normal transverse wave fronts i.e. on the initial pressure and composition of the mixture, can either occur at the inner or outer wall[11]. Thus, more severe damage at the convex wall of a pipeline may well indicate the occurrence of a detonation.

In general, the design of pipelines is based on a static-equivalent containment pressure. Frequently simplifying assumptions are used to relate deformations produced by static and rapidly-applied loads. For instance, the deformation produced by an explosive load, the duration of which is long compared with the response time of the structure, is generally taken to be twice that produced by a static load[13]. In view of the non-uniformity of the distribution of pressure developed behind a blast wave in a sharp bend such assumptions appear to be dangerous over-simplifications, and a complete analysis of the response of all portions of the bend under the spatially-varying loads imposed is necessary for bends with $R/r \leq 3$. Ideally bends in pipelines which may be subjected to internal blast-waves should have $R/r \geq 12$, in order to minimize the effects of the non-uniformity in the distribution of internal stresses.

It should be noted that all shocks in the range $3.1 < M_s < 1.2$ suffered severe attenuation in all the bends tested. The weaker shocks $M_s < 3$ continued to decay during their passage along the exit channel, tending to steep compression fronts at stations 7 and 8, approximately 50 diameters from the exit to the bend. However, stronger shocks $M_o \sim 3$ decayed to a stable velocity at approximately 15 diameters from the end of the bend. The results from $M_s \sim 3$ shocks in bends 2 and 3 suggest that attenuation is less in shallow bends.

Viscous drag may contribute towards retardation of the front. In this case, the longer distances associated with shallower bends should result in greater attenuation. However, this effect will be reduced by the greater periphery of the distorted fronts in sharper bends. Certainly there is no indication of viscous losses in the exit channel itself. Thus, it appears that the overall attenuation within the bend is a result of the complex and non-symmetrical interactions of the subsidiary wave systems induced at the convex and concave surfaces.

5.   CONCLUSIONS

1.   The internal stresses imposed by shocks propagating through bends with $R \sim r$ are highly asymmetric with respect to the cross-section of the channel. Thus, considerable caution must be observed in the design of such a bend.

2.   Internal stresses developed by shocks in channels with bends of $R \geq 10r$ are much less complex, as the shock tends to propagate around the bend at constant angular velocity and minimize its deviations from planarity.

3.   The overall effect of a bend is an attenuation of the shock pressure at ≤ 15 diameters downstream.

6. REFERENCES

1. Nettleton, M.A., Shock attenuation in a 'gradual' area expansion, J. Fluid Mech., 60, 209, 1973

2. Sloan, S.A. and Nettleton, M.A., A model for the decay of the wall shock in a large and abrupt area change, J. Fluid Mech., 88, 259 1978

3. Sloan, S.A. and Nettleton, M.A., The propagation of weak shock waves through junctions, 'Shock Tube Research - Proc. 8th Int. Shock Tube Sym.', Eds. Stollery, J.L., Gaydon, A.G. and Owen, P.R., Chapman and Hall, London, 1971

4. Whitham, G.B., 'Linear and Nonlinear Waves', John Wiley and Sons, New York, 1974

5. Chisnell, R.F., The motion of a shock wave in a channel with applications to cylindrical and spherical shock waves, J. Fluid Mech., 2, 286, 1957

6. Hide, R. and Millar, W., A preliminary investigation of shocks in a curved channel, AERE GP/R 1918, 1956

7. Milton, B.E. and Archer, R.D., Pressure and temperature rise in a logarithmic spiral contraction, 'Shock Tube Research', Proc. of 8th Int. Shock Tube Symp., Eds. Stollery, J.L., Gaydon, A.G. and Owen, P.R., Chapman and Hall, London, 1971

8. Wu, J.H.T. and Ostrowski, P.P., Shock propagation in a 90° bend, Can. Aero. and Space Jour., 22, 230, 1976

9. Takayama, K., Shock propagation along 90° bends, 'Shock Tube and Shock Wave Research' - Proc. 11th Int. Symp. on Shock Tubes and Waves, Eds. Ahlborn, B., Hertzberg, A. and Russell, D., University of Washington Press, Seattle 1978

10. Edwards, D.H., Fearnley, P., Thomas, G.O. and Nettleton, M.A., 1981, Shocks and detonations in channels with 90° bends, 1st Specialists Meeting (Int.) Comb. Inst., 431, Section Francaise du Comb. Inst.

11. Edwards, D.H., Thomas, G.O. and Nettleton, M.A., 1981, The diffraction of detonation waves in channels with 90° bends, To be published in Arch. Comb.

12. Skews, B.W., The shape of a diffracting shock wave, J. Fluid Mech., 29, 297, 1967

13. Edwards, D.H., Fearnley, P. and Nettleton, M.A., 1982, The diffraction of shock-waves in channels with 90° bends, Submitted for publication

ACKNOWLEDGEMENTS

The work was carried out at the Central Electricity Research Laboratories and is published by permission of the Central Electricity Generating Board.

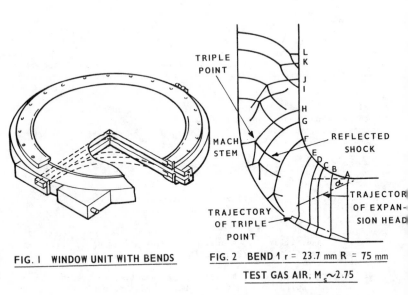

FIG. I   WINDOW UNIT WITH BENDS

FIG. 2   BEND 1 r = 23.7 mm R = 75 mm

TEST GAS AIR, M$_s$ ~2.75

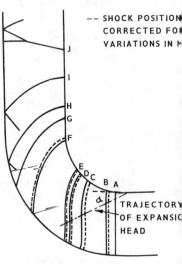

FIG. 3   BEND 1 r = 23.7 mm, R = 75 mm,

TEST GAS AIR, M$_s$ ~ 1.48

FIG. 4   BEND 1 r = 23.7 mm, R = 75 mm

TEST GAS AIR, M$_s$ ~1.22

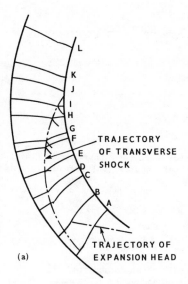

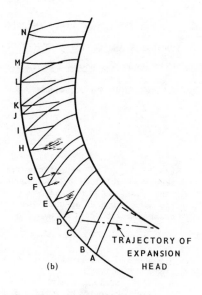

FIG. 5   BEND 2   r = 23.7 mm, R = 150 mm
TEST GAS AIR $M_s \sim 3.02$

FIG. 6   BEND 2, r = 23.7 mm
R = 150 mm **TEST GAS AIR** $M_s \sim 1.20$

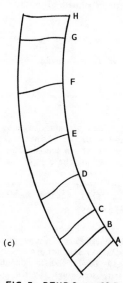

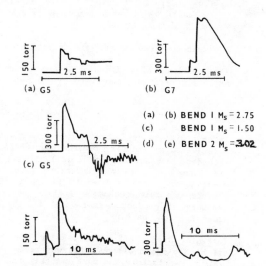

(a) G5

(b) G7

(c) G5

(a) (b) BEND 1 M_s = 2.75
(c)   BEND 1 M_s = 1.50
(d) (e) BEND 2 M_s = 3.02

(d) G5

(e) G7

FIG. 7   BEND 3 r = 23.7 mm
R = 300 mm.   **TEST GAS**
AIR $M_s \sim 1.20$

FIG. 8   TYPICAL PRESSURE HISTORIES DOWN-
STREAM OF SHARP AND SHALLOW BENDS

IGNITION/EXPLOSION RISKS IN FLUIDISED DRYERS

AVOIDANCE AND CONTAINMENT (PRACTICAL ASPECTS - A USER'S VIEWPOINT)

R.W. GRAFTON, Ph.D., C.Eng., F.I.Chem.E.

SUMMARY

In the light of experience the author concludes that batch fluidised dryers as utilised by the pharmaceutical industry only present signifi-cant risks of ignition when drying off highly inflammable solvents. Provided simple precautions are taken this type of drying may be practised with confidence.

Forced draught is preferred to induced draught.

Precautions recommended to avoid risk of ignition include:-

(i)   Stringent earth bonding of all metal components.
(ii)   Discounting difficulties associated with non-conductive fabric filters.
(iii)   Where practicable, choosing drying air temperature to avoid flammable environments.
(iv)   Care in "turning" the bed manually part-way through the cycle to avoid ready discharge of static from the bed of solid being dried.

A simple low-cost proven fire detection/suppression system is recommended

The expediency of positioning dryers on outside walls to provide unimpeded explosion relief is commended.

INTRODUCTION

Batch fluidised drying of powders has been practised to advantage by the pharmaceutical industry for the past three decades or so - initially in secondary manufacture, particularly as part of granulation for tabletting, and later in primary manufacture, i.e. organic synthesis of active substances.   Apart from significant economic and product quality advantages, this technique enabled substantial reduction of risk of exposure of production personnel to physiologically active agents as arose from the earlier common practice of de-traying.

Initially drying off water was involved, but the attraction of drying off highly flammable solvents was recognised, initially in the case of primary manufacture and more recently in secondary manufacture.   In the author's experience ignition has not presented itself as a practical risk until involving highly flammable solvents, although it is appre-ciated that all the organic solid powders are combustible.

Until recently all ignitions have been associated with non-polar solvents, such as heptane, but in 1980 ignitions involving polar solvents, e.g. isopropanol, occurred in at least two fluidised dryers within the author's experience.

An excellent treatise on the general hazards and precautions relevant to the drying of particulates was published in 1977 by the Institution of Chemical Engineers - "User Guide to Fire and Explosion Hazards in the Drying of Particulate Materials". The object of this paper is not to duplicate this comprehensive booklet but rather to explain experience gained within the pharmaceutical industry and to indicate some of the simple precautions which may be applied - not only to reduce the risk of ignition but also to minimise the consequences if and when ignition occurs.

GENERAL DESCRIPTION OF DRYERS
(FORCED AND INDUCED DRAUGHT)

A transportable skip or bowl into which the solid to be dried is put is locked into the dryer body so that heated air may be blown through the solid in the bowl. A fabric filter is hung above the bowl to prevent loss of product. The dryer body, bowl and filter support are of metal and hence highly electrically conductive. The fabric filter conversely is generally an electrical insulator. The dryers are normally located on an outside wall to provide minimum resistance of the explosion relief.

The dryer may be either forced or induced draught. The latter type is illustrated viz:-

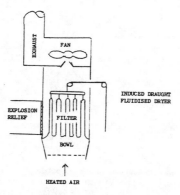

When introducing fluidised drying for primary manufacture the author favoured forced draught systems so as to preclude any risk of ignition arising from a mechanical defect in the fan and to facilitate pressure relief in the event of ignition. It will be appreciated from the above sketch that in the case of an induced draught dryer the fan significantly impedes the value of the exhaust duct for pressure venting.

Furthermore if the fan is housed within the body of the dryer, as has been common practise with induced systems, it is not convenient to locate an emergency damper to isolate the fan from the dryer. Hence in the event of ignition leading to a loss of the fabric filter, the fan is likely to draw fluidised solid into the confines of the fan body, which is not protected by an explosion relief.

IGNITION RISKS

Thermal Instability

The substances to be dried, being destined for therapeutic uses, are unlikely to be so chemically unstable as to give rise to spontaneous heating and ignition. However, it is always sensible to be reassured and a simple check of thermal stability as recommended in the Association of the British Pharmaceutical Industry's publication (1982) - "Chemical Reaction Hazard Analysis" - is favoured.

Mechanical

With induced draught dryers, fouling of the fan casing by the rotor would present a real risk, particularly when evaporating off highly flammable solvents. However the author is unaware of an ignition having arisen from this cause.

Adventitious inclusion of tramp metal in the dryer bowl would similarly present a significant risk. However the very nature of the pharmaceutical industry virtually excludes such gross contamination as a real risk.

Electrostatic

Arising from fluidising dry particulate substances, the generation of surface electrostatic charges is a reality and no doubt explains all ignitions in fluidised dryers within the author's experience.

The energy of a discharge, hence its ability to prove incendive, depends on the nature of the surface. Any isolated electrical conductor, e.g. metal component, must present great concern. Any non-conducting components of the dryer, e.g. a simple fabric filter, will be less likely to lead to a high energy discharge. The particulate matter again being of limited surface area and non-conducting, is considered to be of a low order of risk.

Whether a discharge proves incendive also depends on its environment. It is generally appreciated that the minimum ignition energy for combustible dust clouds is several orders of magnitude greater than for flammable vapour/air mixtures. Hence it is reasonable to suppose that dryers handling highly flammable solvents are likely to be more prone to ignition than dryers handling non-flammable solvents.

Fortunately, in practise, when highly flammable solvents are present in substantial quantities as to give an environment above the lower explosive limit the solid will be "wet" and have a lesser tendency to generate static than when "dry" and when the solvent is substantially eliminated.

What few ignitions have occurred have generally been well into the drying cycle after fluidising has been temporarily interrupted to manually "turn" the bed so as to break up any conglomerates occluding solvent due to imperfect fluidisation. On re-starting the dryer there will be a sudden increased release of solvent at a time of ideal conditions for generation of static, so explaining why ignitions involving highly flammable solvents are more common at this stage of the cycle.

REDUCTION OF RISK OF IGNITION

Inerting

Undoubtedly operating with an inert drying gas would eliminate all risks of ignition. However as dryers can readily be operated with confidence that dangerous occurrences can be contained, if not totally eliminated, the inconvenience and cost penalties are not considered warranted.

Electrostatics

It is considered essential good husbandry to ensure, as far as is practicable, that all metal components are effectively bonded to earth. This is relatively simple in the case of the dryer bowl, where "earth proving" may readily be interlocked with the dryer controls so prohibiting fluidising in the absence of earthing. However it is not so practical to adopt such protective monitoring for the fabric filter support ring and shaker gear. As this equipment is not continually disturbed between drying batches, carefully supervised maintenance and monitoring is favoured instead.

The "non-conducting" fabric filter has presented many operators with much concern. In the author's opinion and experience the concern is unwarranted since the likelihood of an electrically insulating surface producing an incendive discharge, even when drying off highly flammable solvents, is extremely remote. It is far better to accept this philosophy than to endeavour to introduce some means for dissipating the surface charges on the filter, which may give rise to a loss of earth bonding of an electrically conducting component.

A number of different attempts have been made to safely dissipate electrostatic charging of the fabric filter, viz:-

(i)   Incorporating carbon filaments in the fabric. This idea failed at the outset of trials as a result of carbon contaminating the product being dried.

(ii)   Incorporating metal filaments in the fabric. A serious trial using stainless steel was made contrary to the author's advice and led to the first ignition experienced by the author. The solvent at the time was heptane and a major fire ensued. This was before fire detection and suppression had been fitted and the continued running of the forced draught fan resulted in the batch of product effectively burning on a "blacksmith's hearth".

The explanation of the incident is that continual flexing of the stainless steel filaments had led to fatigue giving rise to isolated electrical conductors, the universally recognised worst situation when there is a risk of electrostatic charging.

(iii)   Use of "Epitropic Fibres". Elis, V.S., presented a paper at the 44th Annual Conference of the Textile Research Institute, New York, 1974, describing how carbon (<5 μ) may be permanently incorporated when generating artificial textile fibres, so producing a "conductive" fabric. At least one U.K. operator of fluidised dryers specifies such fabric.

In the author's experience the conductive properties of the fabric incorporating "epitropic fibres" is soon lost when laundering the fabric filters. For this reason the author has not favoured such fabrics, especially as normal non-conducting fabrics are not considered to introduce a significant risk of ignition.

(iv)   Humidification of drying air. It is well appreciated that electrostatic charges on non-conducting surfaces more readily leak away in relatively humid air. One operator known to the author deliberately introduces steam with this aim in view (it should be appreciated that, notwithstanding, this particular operator, who also employs "epitropic fabric", has experienced ignition).   The author does not favour this idea for the presence of avoidably high humidities could be detrimental to the product.

There is also the problem of the solid bed retaining an electrostatic charge at the time it is withdrawn from the dryer body for "turning". Whilst the bowl and operator may be bonded to earth the non-conducting solid presents a problem. Certainly "kicks" have been experienced by operators. Any risk must be confined to a vapour ignition as there is no dust cloud at this stage. The author's recommendation where this is a concern is to allow the bed to stand for a significant period (minutes) to allow the solid charge to essentially dissipate and for the operator to wear a non-conducting glove and use a non-conducting scoop, such as polished hard wood.   Any attempt to employ an earthed conducting implement would be viewed with alarm.

Avoiding Flammable Environments

When drying off highly flammable solvents, it may be practical by temporarily suppressing the drying air temperature to avoid exceeding the lower explosive limit. As indicated above the most likely beneficial stage of the cycle to do this is when the drying is interrupted to "turn" the bed, although caution at the beginning of drying must be expedient.

Where this technique is practical, it is recommended that the drying air be unheated for a period at the start of drying, the heating to be withdrawn sufficiently to ensure the air and bed are cool before interrupting to "turn" and to restart with unheated air.

Whether the technique is of value and what temperatures are relevant will depend on the solvent involved. A theoretical assessment has been made of the internal dryer environment (as % LEL - lower explosive limit) and the adiabatic bed temperature for a number of in-use solvents, based on "wet bulb theory" and the Chilton-Colburn analogy as described by Sherwood, T.K., "Absorption and Extraction" - McGraw Hill. Representing the drying air temperature as "tg" ^{o}C, the adiabatic bed temperature as "ti" ^{o}C and expressing the dryer environment as % LEL, the three parameters are interrelated viz:-

$$tg = A \log \% LEL - B$$

$$ti = C tg - D$$

Where:-

Solvent	A	B	C	D
Methanol	150	210	0.32	15
Ethanol	98	170	0.46	4.8
Isopropanol	73	126	0.51	0.83
n-Hexane	130	340	0.51	7.1
n-Heptane	84	190	0.70	1.4
n-Octane	56	92	0.88	0.65
Toluene	65	120	0.73	0.56
m-Xylene	57	89	0.86	-0.93

This assessment predicts that for n-heptane, n-hexane and toluene it is virtually impracticable to suppress evaporation to below the LEL (in the case of n-hexane much of the drying off will be at above the UEL - upper explosive limit).

The significance of the adiabatic bed temperature is to indicate the risk of freezing out water from the drying air. Methanol presents such a risk if attempting to operate with "cool" drying air, but fortunately heated drying air in the case of this solvent is unlikely to give a dryer environment in excess of the LEL.

Below is shown the predicted relationship between drying air temperature and maximum dryer environment conditions for isopropanol. It should be noted that $65^{\circ}C$ is a common drying temperature which suggests that when sufficient solvent is present, a flammable environment is likely.

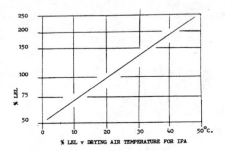

% LEL v DRYING AIR TEMPERATURE FOR IPA

The following sketch is an actual assessment of the dryer environment when drying off isopropanol. For a significant period a flammable environment exists.

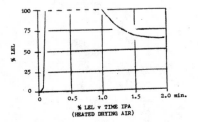

% LEL v TIME IPA
(HEATED DRYING AIR)

In the light of the guidance from the theoretical assessment the sketch below describes the practical conditions with a period of unheated air to remove the bulk of the solvent.

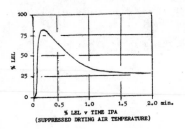

% LEL v TIME IPA
(SUPPRESSED DRYING AIR TEMPERATURE)

Thus where the intrinsic characteristics of the solvent permit, there is clear advantage in suppressing the drying temperature until the bulk of the solvent is removed. Reminder is made of the value of cooling before interruption and "turning". In the case of such solvents as n-hexane, where operation below the LEL is impractical, high drying air temperatures are attractive to ensure the environment is above the UEL as far as is practicable and the transition to below the LEL is as short as possible.

CONTAINMENT OF CONSEQUENCES OF IGNITION

No matter how vigilant operation is, there is always risk of human error leading to an absence of a vital precaution, such as a loss of earth bonding of a metal component, particularly in the filter bag support and shaker gear. Thus for the same reason that sprinkler systems are fitted in no-smoking areas, some provision to minimise the consequence of ignition is warranted.

In the light of experience, the author's considered understanding of the sequence of ignition in pharmaceutical fluidised dryers is that primary ignition always involves a flammable solvent/air mixture and not a dust cloud. This can then ignite a dust cloud with explosive force, particularly in the case of induced draught systems after destruction of the filter permits a dust cloud to be sucked into the fan casing. This two stage ignition, the first stage being relatively slow, is supported by experience gained with the simple fire detection/suppression system described below.

So successful has this system proved that ignition has occurred without damage to the equipment (save partial rupturing of the explosion relief panel and charring of the fabric filter) and without loss of product. On one occasion the explosion relief panel was pulled inwards, rather than being punched out.

This simple idea has proved to be extremely reliable and inexpensive, as would not be anticipated with an explosion suppression system.

The sketch shows the basic system. Carbon dioxide at 50 Bar pressure is coupled to a stainless steel manifold. V1 is a stainless steel ball valve held closed by a pneumatic actuator fed with air at 5.6 Bar through the bursting tube. A fire in the dryer causes the polythene tube to melt (at 90°C) thus releasing the air pressure and opening V1 and flooding the drier with cardon dioxide. V2 is a stainless ball valve, manually operated, fitted with a limit switch S1 which is only operated when the valve is fully open. When the drier is in use V2 is open and thus S1 is actuated feeding air to S2 which feeds air from the bursting tube to the pressure switch thus allowing the fan to be run. In the event of a fire the pressure switch opens thus stopping the fan and so stopping air being fed to the fire. In addition S3 switches and operates a cylinder which closes a damper cutting off air flow from the fan.

During cleaning and maintenance operations when inadvertent breakage of the bursting tube might prove dangerous, V2 is closed to isolate the carbon dioxide. The fan cannot be run until V2 has been fully re-opened. An additional pressure switch can be fitted to give automatic fire alarm operation.

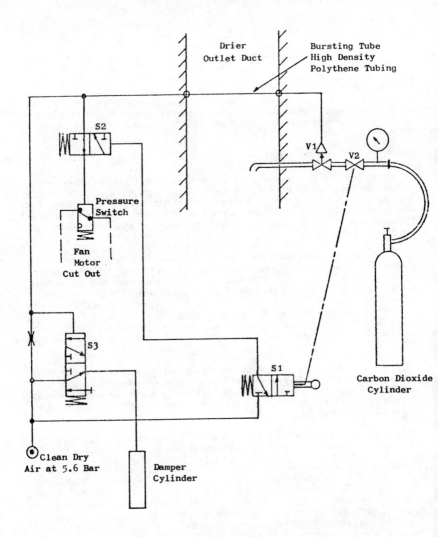

STATIC CHARGING IN POWDER CONVEYING SYSTEMS

DOUGLAS H. NAPIER

Department of Chemical Engineering and Applied Chemistry
University of Toronto.

SYNOPSIS

Measurements have been made of static charging of 4
particulate materials conveyed pneumatically under low
density conditions. The materials were corn dust, polystyrene,
granulated sugar and wheat flour. The total charge, Q_T, on the
materials after conveying through tubes 1m in length was given
by:

$$Q_T = Kt V^{1.7} D$$

where t is the collection period and
where V is the conveying velocity and D is the diameter of
the tube. It is shown that this relation is consistent with
a model of charging based on the concept of dielectric work
function.
Measurements were also made on two industrial systems viz.
a grain terminal and a bulk flour carrier. The results in
both cases are in satisfactory agreement with observations
and measurements made in the laboratory.

INTRODUCTION

Charge separation at interfaces and consequent residual
charges produced in the separate phases are both of intrinsic
interest and of practical importance. The latter aspect
involves the explosion protection of installations wherein
combustible dusts are handled and transferred. The requirement for guidance in design and for estimation of hazards
arises. Another practical aspect is that of the adherence
of dusts to surfaces and the free-flowing properties of
powders.

In the conveying of powders multiple contacts are inherent
in the operation. Such contacts take place between particles
and containing walls, between particles and conveyors and on
a particle-to-particle basis. The multiplicity of contacts
results in numerous opportunities for movement of charge
arising from equalisation of energy levels.

In this paper static charging will be considered   from the
point-of-view of parameters in a system that affect the amount
of charge separated. This quantity may be a factor in the
initiation of explosion. The requirements for the ignition
of dispersions of combustible dusts leading to explosion, is
often expressed in terms of minima viz:

        explosible concentration
        oxygen concentration
        ignition energy, ignition temperature

Concern here is with energy of ignition. An important aspect of the matter is the presence of and necessity of eliminating vessels, connections and process tools that may be electrically isolated (1). Extra emphasis has been drawn to these situations by the fact that some of the earlier values quoted for energy of ignition of dust dispersions indicated that static charging was not a highly probable source of ignition. More recent work by Eckhoff (2) and Napier et al (3) has shown that ignition of dust clouds by sparks from a capacitative circuit can be achieved at much lower spark energies. Some examples are given in Table I; new values are usually <10% of the total circuit energy.

TABLE I Typical values of Minimum Ignition Energy

Material	Average value $_{mJ}$	New value
Aluminium flake	10	1(1)
Wheat dust	50 – 60	20(1)
Ground sulphur	15	0.3(1)
Polystyrene	40	0.3(2)
Cornstarch	60	0.3(2)
Polyethylene	20	3.5(2)

The work described below is divided into two main parts. These are laboratory investigations and measurements in industrial systems. These parts are comprised of
    (i) a laboratory rig wherein particulates (corn dust, polystyrene and sugar) were pneumatically conveyed under low density conditions. A further set of measurements was made with wheat flour in a similar rig.
    (ii) an industrial-scale grain-conveying unit
        a bulk flour carrier during unloading.
In each case measurements of static charge were made; these are reported below. Their significance can be viewed in several ways. Thus:
    the magnitude of the separated charge is of importance per se
    the effect of some of the parameters in the system has been demonstrated
    some correlations have been sought

INVESTIGATIONAL WORK

Laboratory Rig for Pneumatic Conveying of Powders

This system was operated under low density conditions (i.e. less than 5% mass per unit volume).

Apparatus
The apparatus is shown diagrammatically in Figure 1. Two arrangements for obtaining a steady feed rate of particles were developed. These were:
    (i) a star-feeder
    (ii) a 6-start screw mounted inside a close-fitting, smooth bore housing.
The rate of both feeds was controlled with a triac transformer. The second arrangement was particularly successful with fine powders such as wheat flour. Difficulties arose with granulated sugar with which the star-feeder performed well.

In that each of the materials used in the experiments, was an "as-received" sample there was a range of particle sizes present in each of them. An estimate of the relative fineness is given by the approximate mean particle size. These were as follows (in μm):

polystyrene	200
corn dust	50
granulated sugar	500
wheat flour	range 20 - 60

The materials were conveyed through glass tubes(and in the case of wheat flour also through a copper tube) using nitrogen with linear conveying velocity in the range 3.5 to 12m s^{-1}. The concentration of powder was held within the range 0.01 to 5% (mass/volume). At low velocities the conveying of the dust was observed to be uneven, as was the effect with granulated sugar in the velocity range used here. The range of tubes through which the dusts were conveyed were of the following diameters 0.5, 0.75, 1, 1.25 and 1.5cm. The length of the tubes was 1m and for wheat flour 1.3m.

At the downstream end of the tube the dusts were separated from nitrogen used for conveying. The charge on the particles was measured by collecting them in a vessel of known capacitance and observing the rise in voltage for a measured flow of material. This measurement and direct charge measurements were made with a Keithley 610C Electrometer that was connected directly to the collector.

Results of Experiments with Polystyrene, Corn Dust and Granulated Sugar

The results are given in Figure 2(polystyrene), 3 (corn dust) and 4 (granulated sugar). There was appreciable scatter of measurements made with granulated sugar in the system. The values shown in Figure 4 indicate the trends of the variation and give an estimate of the absolute values. Over the range of conveying velocities examined the total charge collected increased by a factor of up to about 20 times and the charge density rose by up to 7 times. These increases were obtained for a maximum increase in conveying velocity of 5 times.

In the experiments with polystyrene the loading (mass per unit volume) was held constant at 0.1%. The total amount of charge collected increased with tube diameter. This would be expected in view of the larger amount of powder collected in unit time. Further, in view of the fact that the transport was at lower density opportunity for the particles to strike the wall of the tube was unlikely to be restricted. Particle-to-particle collisions should be at a minimum.

In the experiments with corn dust (generated during the mechanical handling of corn) the loading was held constant at 0.05%. Similar results to those obtained with polystyrene were obtained. The variation of charge density (expressed as C kg^{-1}) with tube diameter and with conveying velocity is shown in Figure 3.

With both materials charge density decreased with tube diameter. This may be set down qualitatively as due to another diameter-dependent factor viz the frequency of collision of particles with the tube wall. Further in view of the fact that both quantity of dust and surface area of the tube increase with the square of the tube diameter, it is note-worthy that the total charge varies directly with the diameter and not as a function of it.

In fact for reasons given in the next section the total charge $Q_T(C)$ can be predicted from an experimental relation of the form

$$Q_T = KtV^{1.7} D$$

where V is the velocity (m s^{-1}) Q_T is charge collected (C) in
D is tube diameter (m)   t   time (s)

For samples conveyed under similar conditions in these experiments, K has values as follows

polystyrene 7.8×10^{-9}
corn dust 9.3×10^{-9}

The nature of K will be explored below.
Two samples of sugar were used. That designated in Figure 4 as fresh was taken directly from the drying plant at a sugar refinery and that designated conditioned had been stored for 4 weeks. The fresh sugar acquired a greater charge when passed through the apparatus than the conditioned sugar. Both carried charge in excess of 10^{-5}C kg^{-1}. As for polystyrene and corn dust the build-up of charge was greater in tubes of smaller diameter.

In later experiments using the 6start screw feeder, measure-ments of charge on sugar passed through a glass tube of 1.5cm diameter at a conveying velocity of 7m s^{-1} at voidages ranging from 0.14 to 0.37% were made. These were negative and in the range of 0.3 to 1.8 x 10^{-5}C kg^{-1}. Values of K were in the range 10 - 20 x 10^{-9}. Reproducibility in the laboratory was not good due to irregularities in feed and flow.

Factors in Prediction of Amount of Charge

Charging may be considered on the basis of equalisation of electronic energy levels and therefore of flow of electrons at collisions between materials. Particle-to-particle collisions will be ignored here because of the low density condition. It may be shown using available data on turbulent flow in pipe that the number of impacts, n, of a single particle in a turbulent stream is given by:

$$n = \frac{L}{16D}$$

where L is the length of the pipe and D its diameter
Hence for i particles the total number of impacts N is:

$$in = \frac{iL}{16D} = N$$

If voidage of the suspension is Ω time interval is   t
particle radius   r   linear gas velocity   V

then, $i = \dfrac{3\Omega D^2 \, tV}{16r^3}$ and $N = \dfrac{3L\,\Omega\,DtV}{256r^3}$

The charge transfer per particle-to-wall collision, q_{21}, may be expressed as $q_{21} = \text{constant} \in \dfrac{(x_1 - x_2)}{\lambda}$

where Ω is the permittivity

χ_1 and χ_2 are dielectric work functions

λ is the depth of penetration of transferred charge

q_{21} is in the units of charge per unit area (C m^{-2}). The area of contact per collision q_{21} has been taken in terms of the relation due to Hertz

$$a_{21} = 1.1r^2 \left(\frac{\pi P}{2E}\right)^{\frac{1}{3}}$$

where E is Young's Modulus

and P $\frac{1}{2}V^2 \rho_g$ in which ρ_g is the density of the gas

Whence $a_{21} = 1.1r^2 \left(\frac{\pi P}{4E}\right)^{\frac{1}{3}} V^{\frac{2}{3}}$

The total charge, Q_T, obtained in the time t is given by

$$Q_T = N q_{21} a_{21}$$

Whence $Q_T = KtV^{5/3} D$

in which K embraces the constants in the system i.e. both apparatus, L and material. Davies (4) determined q_{21} for the glass/polystyrene system as 4 x 10^{-6}. His experiments were with clean surfaces in vacuo. If K is calculated on the basis of this value it is equal to 15 x 10^{-9} which is not far removed from the value obtained here of 7.8 x 10^{-9}.

$Q_T = N q_{21} a_{21}$

$$= \left[\frac{3L \Omega DtV}{256 r^3}\right] \times \left[constant \in \frac{(\chi_1 - \chi_2)}{\lambda}\right] \times \left[1.1 r^2 \left(\frac{\pi \rho_g}{4E}\right)^{\frac{1}{3}} V^{2/3}\right]$$

Whence $Q_T = KtV^{5/3} D$

Thus $K \propto \frac{L \Omega}{r} \left(\frac{\rho_g}{E}\right)^{\frac{1}{3}} \in \frac{(\chi_1 - \chi_2)}{\lambda}$

A full pronouncement upon K calls for much more extensive study. Proportionality to L would be expected under the conditions used; the onset of $(Q_T)_\infty$ was not explored. The effect of Ω is shown in Figure 5, but this is for one material only and over a limited range of voidage. Further, it seems likely that the effect of r will change markedly over a wide range of particle size.

Values of charge density (Ckg^{-1}) can be deduced from total charge if velocity of conveying, tube radius, time of collection and voidage are known. Thus with the same sample of a material conveyed under conditions 1 and 2 the following proportional relationship holds at the same voidage

$$C.D.' = Q_T' \times \frac{4}{\pi D'^2 V' t' \Omega}$$

and

$$\frac{C.D.'}{C.D.''} = \frac{Q_T'}{Q_T''} \times \frac{D''^2 V'' t''}{D'^2 V' t'}$$

E57

In the experimental work reported here t' and t'' were equal (15s) and the voidage (Ω) was held constant for each material.

Results of Experimental Work with Wheat Flour

Measurements with wheat flour were made in the same type of apparatus as that shown in Figure 1 using the 6start screw feeder. However, some of the parameters in the system were examined in relative isolation.

Effect of voidage. Variation of charge density with voidage is shown in Figure 5, direct proportionality was observed over the range examined. The regression line gave

$$10^6 \text{ x C.D.} = 72.6 - 15.21 \Omega$$

As voidage decreases the probability of particle-to-wall collision increases and therefore the probability of particle-to-particle collision decreases. The charge generated would be expected to increase with decreasing voidage.

Effect of tube diameter. The total charge residual on the collected solids increased with increasing diameter as shown in Figure 6; the measurements were made at two values of relative humidity (R.H.). Charge collected was greater at the lower value of R.H. (65%). Charge density (C kg^{-1}) decreased with increasing tube diameter over the range examined.

When the material of the tube was changed from glass to copper a decrease in charge was noted. No significant difference was registered between grounded and ungrounded earthed copper tubes These results are shown in Table II.

TABLE II Effect of tube material on charging

Type of tube	Charge density (C kg^{-1} x 10^5)
Copper, grounded	3
Copper, ungrounded	3.2
Glass	4.7

Effect of conveying velocity. The results obtained from a short range of values of conveying velocities is shown in Table III; the voidages were not constant from one run to the next, so for purposes of comparison the ratio of Q_m/Ω was used. In the light of the foregoing model this quantity was divided by $V^{5/3}$. It will be noted that the final ratio exhibits an acceptable degree of constancy, which suggests that the five thirds power of velocity also applies to charging of flour. The glass tube used in these experiments was 1cm in diameter.

TABLE III Effect of conveying velocity on charging of flour

Gas velocity (V) m s^{-1}	Q_T x 10^6C	Ω	$\dfrac{Q_T}{\Omega\, V^{5/3}}$
4.2	6.9	0.7	0.9
5.1	10.3	0.8	0.8
6.6	10.8	0.4	1.1
6.8	19.6	0.6	1.2
10.3	19.8	0.4	0.9

All of the charges collected on wheat flour were positive as were those on corn dust. Charges on polystyrene collected after conveying were negative.

Industrial-scale Grain Conveying Unit

Corn was handled, stored and distributed at the terminal where the measurements reported here were made. The corn was lifted from bulk carriers by suction and by using bucket elevators and conveyor belts it was transferred to to silos. From the silos the corn was transferred, for distribution, into trucks rail cars, the holds of small ships and barges, again using bucket elevators and conveyors.

Dust was generated in handling the corn and measurements were made at several situations where it was dispersed. For this purpose a Davenport electrostatic field meter (of the mill type) was used and field strength was recorded.

(i) Spout into a barge. Velocities of the corn in the down comer to the spout were in the range 5 to 8m s^{-1}. A dust cloud, hemispherical in shape and of estimated radius 10m, was formed at the spout. The field strength measured 10m from the centre of the cloud was 10kVm^{-1}.
From this measurement it may be deduced that:
the charge density was ca 2.7 x 10^{-8}C m^{-3}
and since the volume of the cloud was 2095m^3 the total charge in the cloud was 5.6 x 10^{-5}C. If in order to obtain an estimate of the energy in the cloud the assumption is made that the field is uniform, the result is of the order of 1J. There is a clear possibility of hazard from conductors isolated from ground and arrayed unfavourably with respect to the dust cloud.

(ii) Belt conveyor. The belt conveyor at which measurements were taken was about a metre in width and the field mill was centrally disposed 2.5cm above the conveyors. The field strength was 1.5kVm^{-1}. If the belt is treated as long with respect to its width, the charge=7.4 x 10^{-9}C m^{-2}

In view of the dusty nature of the corn a shallow dust cloud formed above the conveyor. To limit the spread of this dust cloud canvas curtaining enclosed the conveyor. This limited the dust cloud in the building housing the conveyor to a thin haze in appearance. The charge density of the shallow cloud was 3.5 x 10^{-1} C m^{-3}.

(iii) Storage bin. The bin was in active work so that its head space was filled with a dust dispersion. The corn had been handled in the manner outlined above in order to load it into the bin. Measurements of field strength were made at distance of 1,2 and 3m below the top of the bin. The values were 3.3, 5 and 7.3 kV m^{-1}. It was estimated that the value at the grain surface was 17.5 kVm^{-1}. This leads to an estimate of charge per unit area equivalent to 8.5 x 10^{-8}C m^{-2}.

It has been shown that energy may be present in considerable quantities in dust clouds produced by grain handling. This energy can readily be transferred to items of plant, smaller containers and tools and to items of personal equipment. There is a clear requirement for grounding in these latter cases.

It will be noted that the charge density on corn dust in laboratory experiments was greater than that in the grain conveying unit. In the former the dust was conveyed in a gas stream and particle-to-particle collisions were minimised. In the latter case, particle-to-particle collisions preponderated and the nature of movement of material was largely "en masse".

Bulk Flour Carrier During Unloading

A bulk-flour carrier containing slightly under 20 tonnes was unloaded by fluidising the load and passing it out through a hose into a storage silo. The demand for air was up to $12m^3$ min^{-1}. Velocities thereby produced were up to 25ms^{-1} in the transfer hose. The conditions were therefore more severe in terms of velocity than those in laboratory experiments. The situation was less favourable to particle-to-particle collisions than in low density transfer.

The carrier was fitted with 4 measuring positions.
(A) a probe extendable to 1.2m into the tank and mounted at a man-hole
(B) a fixed disc probe mounted at a man-hole and following its contour
(C) field-mill in a man-hole cover
(D) collecting electrode in the discharge hose.

Electrodes (A),(B) and (D) were connected to a Keithley 610C Electrometer which indicated the current drawn.

A steady current of 1.5×10^{-7}A was drawn from position (A) which probed the space above the fluidised flour. The area of the disc at (B) was $0.004m^2$ and a steady current of 1.5×10^{-7}A was also drawn therefrom. The maximum reading from the field-mill at (C) was $0.4kVm^{-1}$. The current drawn from (D) which had an effective area of $0.07m^2$ varied in the range 0.5 to 3.5×10^{-4}A. A simplified approach to charge collection suggests that the charge density of the wheat flour was in the range 0.8×10^{-5} to 5.4×10^{-5}C kg^{-1}. These values are in broad agreement with those determined in laboratory experiments. The handling of material during unloading of the carrier was more conducive to multiple collision than in the handling of corn in the grain terminal.

With these quantities of charge available concern must be felt for the energy of discharges available from the system. This is confirmed by observation of the discharges. Several situations with such a bulk carrier appear to present active hazard. One such situation occurs when unloading has been completed. If the volume of the carrier is taken as ca. $40m^3$ and dispersed dust is at a concentration of 100mg l^{-1}, the total quantity of material may be carrying charge in the range $3 - 18.9 \times 10^{-5}$C. Using Haase's approximation for capacitance(5) the estimate of stored energy in the system is of the order of 1J. Reference (3) has been made to percentage of stored energy released into a discharge. These values were for capacitors in the range 500 to 8000pF. The estimated value of capacitance for the carrier is 450pF; on this basis a discharge of up to ca. 0.1J might be anticipated.

CONCLUSION

Attention has been drawn previously(6) to the charging of
powders and empirical relations have been established for
charge on the solids and various parameters in conveying
systems. This approach and the hazard to which it directs
attention has been reiterated here. Relation is to the former
in laboratory rig and the latter in large scale equipment. It
is suggested that the static charge in pneumatic conveying
may be put on a quantitative basis by invoking a charging model
based on dielectric work function

REFERENCES

1.Napier, D.H. 1971 I.Chem.E.Symp.Ser. No34 p.170
   Napier,D.H. Rossell,D.A. 2nd Internat.Loss Prevention Symp.
      1977 p.157
2.Eckhoff,R.K. 1975 Comb.and Flame 24 53
      1976   ibid   27 129
3.Hay,D.M.,Napier,D.H., Hazards VI 1977 I.Chem.E.Symp.Ser.
      No.49 p.73   Canadian Sectn34-1
   Napier,D.H.,Sinukoff,R.J., 1983 Comb.Inst.Spring Tech.Meet.,
4.Davies D.K.,1967 I.P.P.S. Conf.Ser.No4 P.29
5.Haase,H., Electrostatic Hazards Verlag Chemie New York 1977
6.e.g.Ramacker,L., 1970 E.C.F.E.,1 p.370
      Eden,H.F., 1972 Loss Prevention 6 27.

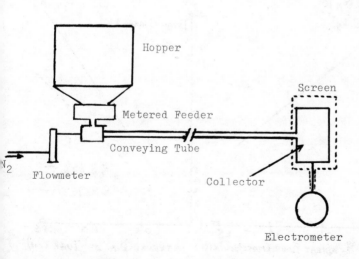

Figure 1. Laboratory Rig for Pneumatic Conveying

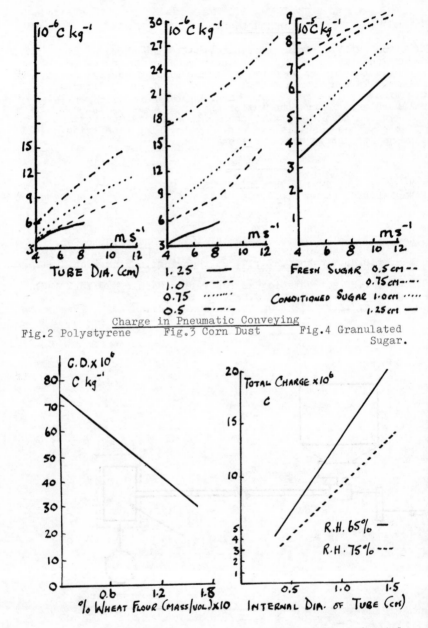

Charge in Pneumatic Conveying

Fig.2 Polystyrene   Fig.3 Corn Dust   Fig.4 Granulated Sugar.

Fig. 5. Variation of Charge Density with Voidage

Fig.6. Variation of Total Charge with Tube Diameter.

Explosion Relief

Chairman

W.B. Howard
Consultant, USA

PERFORMANCE OF LOW INERTIA EXPLOSION RELIEFS FITTED TO A 22m^3 CUBICAL CHAMBER

P F THORNE, Z W ROGOWSKI AND P FIELD

FIRE RESEARCH STATION, BUILDING RESEARCH ESTABLISHMENT

SUMMARY

Explosion tests have been carried out in a 22 m^3 cubical chamber fitted with explosion reliefs with K values in the range 1 to 8 using stoichiometric natural gas, propane and ethylene with air mixtures. The performance of lightweight(plastic film and fibreboard) vent covers in mitigating explosion pressures has been assessed. The nature of high secondary pressure peaks caused by acoustic oscillation has been investigated. The results are compared with previously published data.

INTRODUCTION

The majority of explosion venting experiments have been carried out in relatively small (<2 m^3) vessels or compartments fitted with small ($K > 4$) vents, where K is defined as the ratio of the cross sectional area of the compartment wall containing the vent to the area of the vent (A_c/A_v). Whilst these data are useful for application to plant items, it is useful to have available experimental results for larger volumes and vent areas because scaling laws in general have not been rigorously tested and cannot be expected to predict combustion phenomena which may become apparant only on the larger scale. The present work decribes experiments in a 22 m^3 compartment containing no obstacles with initially quiescent mixtures and vents with lightweight coverings in the range K = 1 to 8.

EXPERIMENTAL DETAILS

The explosion cell facility operated by FRS at its Cardington outstation is shown in fig 1. The complete rig comprises two adjacent chambers, 2.4 m x 2.4 m x 3.7 m (22 m3), interconnected by a doorway with a 1.2 m x 2.4 m interconnecting corridor running along the back and each side. For the present experiments only the left hand compartment was used. The facility is constructed in reinforced steel plate (11 mm). Retractable probes enable gas samples to be withdrawn for analysis, by IR, at various heights. The front face contains the vent which can be varied in size and shape and fitted with a range of vent covering materials. Pressure is measured at selected locations using Kistler gauges type 7031 (inertia compensated); Data is currently acquired digitally at a sampling rate of 2 KHz processed and plotted on a desk top computer; previously FM tape recorders and oscillographs have been used.

EXPERIMENTS CARRIED OUT

The experiments reported here cover natural gas, propane and ethylene in air mixtures at K factors in the range 1 to 8 using 125 /um and 250 /um polyethylene film, 50 /um "Melinex" (polyester) film, and 12.5 mm fibreboard. Details of the combinations tested are shown in Table 1. The

Symbol	Vent material	Thickness	Vent Configuration	Gas	Conc in Air (% v/v)
O	Polyethylene	130 μm	A,B,C	NG	10
△	"	130 μm	A,C	C_3H_8	4.3
▲	"	250 μm	B, C	C_3H_8	4.3
□	"	130 μm	A	C_2H_4	6.25, 6.5
X	Polyester	50 μm	B,C	NG	10
+	"	50 μm	B,C	C_3H_8	4.3
Ø	Fibreboard	12.5 mm	B,D,E	NG	10

Vent	K	Configuration
A	1	
B	2.3	
C	4	
D	4	
E	8	

TABLE 1   Details of vent materials and configurations used

experiments cover a range of burning velocity, vent area ratios (K) and vent covering material characteristics (burst pressure and failure mode).

RESULTS

1. Bursting pressure of the vents

The bursting (opening) pressure of the vents covered with the various materials was found to depend on the area of the vent (K factor) and the gas used i.e the rate of pressure rise prior to venting. In the absence of direct flame impingement, polyethylene was seen to fail due to stretching of the material whilst polyester and fibreboard underwent only limited distortion before failing in a brittle mode. Vent burst pressures for the various materials used are shown in fig 2. These values were obtained for plastic film by recording the instant during the explosion when an electrically conducting strip of thin aluminium foil stretched across the material broke. For fibreboard, timed video records were analysed and the pressure at which cracks in the vent panel were seen to open was taken as the vent burst pressure : in practically all cases this pressure was identical to the first pressure peak. From fig 2 it can be seen that for a K1, 130 /um polyethylene, vent the burst pressure increased as the burning velocity of the gas mixture and hence the rate of pressure rise increased, viz . $C_2 H_4 > C_3 H_8 >$ natural gas.

2. Explosions with K = 1

With all gases tested (Stoichiometric natural gas, propane, ethylene) the first peak (P_1) gives the maximum pressure. This peak is followed in nearly all cases by a negative pulse followed by a number of supplementary but low peaks. With natural gas, after about 1 second from ignition (500 msec from P_1) an oscillation was seen on the pressure record. This was amplitude modulation about zero pressure at a frequency of 192 Hz. The maximum modulation was of the order of $\pm$ 1.5 kN/m^2, about equal to P_1. It is thought that the source of this vibration was acoustic, for a half wavelength equal to the height of the cell, the speed of sound giving rise to this vibration would be 995 m/sec, corresponding to a mean gas temperature of 1743°C ($c \propto \sqrt{T}$). With propane a similar type of oscillation was seen at about 600 ms after ignition. The results for ethylene were not of adequate quality to reveal any oscillation.

High speed (500-1500 fps) ciné records were taken of the flame front in these experiments. The records confirmed that the flame front expanded smoothly and spherically up to the time that the vent ruptured. From the records for propane the burning velocity was calculated to be 0.45 ms^{-1}. The spherical structure of the flame was seen to be disrupted when the vent burst.

The maximum over pressure results for K = 1 with natural gas and propane are shown in figs 7 and 8, together with the only other available result for K = 1 (with stoichiometric propane).[4]

Several correlations of maximum explosion over-pressure attempt to incorporate burning velocity (Su) in order to predict the effect of different gases and gas mixtures and, by the use of an empirical factor on the burning velocity of turbulence. Correlations proportional directly to Su^1 and Su^2[2,3] have been proposed. Results for stoichiometric natural gas and propane and the mixtures of ethylene with air (6.25% and 6.5%) for K = 1 have been plotted against Su in fig 9. This correlation indicates a dependence of maximum overpressure on Su^2.

3. Explosions with K > 1

With K = 2.3 a negative pressure wave followed the first peak but was more pronounced with the more brittle materials (Melinex and fibreboard) than for the ductile materials (polyethylene). The maximum pressure would be determined by either the first peak or the second or a subsequent peak. Typical results are shown schematically in fig 3. Trace A shows the negative pulse which follows P_1.

As vent sizes were reduced (K = 4 and 8) the negative pressure wave referred to was successively abated; see fig 4 which shows results for natural gas with fibreboard at K = 2.3, 4 and 8. This reproduction of actual pressure records indicates that K is increased, the peak which determines the maximum over pressure changes from the first to the last of a succession of secondary peaks. The vibratory nature of this important secondary peak is clearly seen. The vibratory peak is particularly important for the highest value of K tested (K = 8) in which case it is many times the first peak value. The general pattern was also seen with other vent materials and with propane (both at K = 2.3 and 4).

The vibratory peak was seen to occur rather late in an explosion history. In fact it occurred at about the same time as the 'acoustic' peak seen in the case of K = 1. An explosion at K = 8, where the vibratory peak was most severe, was repeated with 76 mm thick mineral fibre wool as a lining to the floor and the three vertical sides, not containing the vent, of the chamber. The mineral wool was expected to damp any acoustic oscillation which was set up.[5] The result is shown in fig 5 in comparison with the undamped explosion.

The vibratory peak mentioned above was seen, for K = 2.3, only for ignition at the geometric centre of the chamber. The effect of alternative ignition locations is shown in fig 6. Where pressure records are reproduced for centre ignition, ignition at the centre of the rear wall and ignition at the centre of the front wall containing the vent. For rear ignition there is a high subsequent peak but not of a vibratory nature; there is some vibration at the end of the record. For ignition at the front, adjacent to the vent, the first peak determines the maximum over pressure, the subsequent pressure being in the form of a decaying oscillation at a frequency of about 21 Hz.

Overall results are shown plotted in figs 7 and 8, for natural gas and propane respectively, in which maximum over pressures are plotted against vent area K. Indication is made, where possible, of the overpressures which are vibratory in nature. Comparison is made with results of other work with enclosures of volume greater than 1 m^3, for natural gas[6,7] in fig 7 and propane[4,8-10] and pentane[11,12] in fig 8. In many of these comparative cases it is not known whether the maximum over pressures reported were as a result of vibratory peaks.

DISCUSSION

This series of experiments has covered a sufficiently wide range of variables to demonstrate that the maximum over pressure obtained in any given explosion is strongly dependent upon the condition leading to the explosion, viz gas mixture, vent area, vent covering material (strength and bursting characteristics), explosion compartment contents and location of ignition source. Correlations of maximum over pressure obtained from experiments with a limited range of variables and conditions cannot be used with confidence to predict expected over pressures in a general sense. Correlations of such sets of data are often satisfactory within themselves but the totally available data taken as a whole is not readily correlatable,[13] although attempts have been made[3].

Theoretical correlations based on simplified combustion models are at present also inadequate because they cannot, for example, predict the complex acoustical interactions leading to high secondary oscillatory pressure peaks.

Vibratory pressures, which can result in high secondary peaks with small (K = 8) vents are an important feature of explosions ignited centrally. Similar phenomena have been reported elsewhere[5,7,14] and it is clear that certain of them must be attributed to acoustic phenomena.

Taylor-Markstein (TM) instabilities have been proposed as being an important acceleratory mechanism in the development of pressure in vented explosion with initially open vents.[8] In the current work high speed (1000 fps) and photography has shown that immediately after vent rupture distortion occurs of the initially spherical flame front in the form of wrinkling similar in appearance to that which has been attributed to TM instability[8] at about the time that pressure vibrations are seen on pressure records the wrinkled flame front oscillates at about the fundamental front to back frequency of the chamber. Clearly the relative roles of and interactions between Rayleigh and TM instabilities is complex and merits further investigation.

ACKNOWLEDGEMENT

This paper forms part of the work of the Fire Research Station, Building Research Establishment, Department of the Environment. It is contributed by permission of the Director Building Research Establishment.

The assistance of I G Buckland and D Henderson in performing the experiments and producing the explosion pressure data is gratefully acknowledged.

REFERENCES

1. Cubbage P A and Simmonds W A. 'The design of explosion relief for industrial drying ovens' Symposium on Chemical Process Hazards Institution of Chemical Engineers 1960 p69.

2. Cubbage P A and Marshall M R. 'Pressure generated by explosions of gas-air mixture in vented explosions'. Inst. Gas Engrs Communication 926, 1973.

3. Bradley D and Mitchesson A. 'The venting of gaseous explosions in spherical vessels II - Theory and experiment' Comb. and Flame 32 (1978) 237-255.

4. Komitten for Explosion fursok, Bromma 1957, Sluttapost. Stockholm (April 1958).

5. Zecuwan J P. 'Current review of research at TNO into gas and dust explosions'. Proc - Int'l Corp. on Fuel-Air Explosions, McGill University Montreal (1981) University of Waterloo Press (1982).

6. Pasmann H J, Groothuizen Th. M, and Gooijer H de. 'Design of pressure relief vents' Loss Prevention and Safety Promotion in the Process Industries. Proc. 1st Int'l Symposium. The Hague (1974).

7. Zalosh R G. 'Gas explosion tests in room-size vented enclosures' Loss Prevention 13 (1979) p98-108.

8. Solberg D M, Pappas J A and Skramstad E, 'Observations of flame instabilities in large scale vented gas explosions' 18th Symp. (Int'l) on Combustion (1981) p 1607-1614.

9. Donat C. Data reported NFPA Code 68 (1978) 'Explosion Venting'.

10. Cotton P E and Cousins E W. 'Design closed vessels to withstand internal explosions' Chem. Eng. 58 (1951) 133.

11. Harris G F P and Briscol P G 'The venting of pentane/air explosions in a large vessel' Comb. Flame 11 (1967) 329.

12. Burgoyne J H and Wilson M J G, 'The relief of pentane/air explosions in vessels' Symp. on Chem. Process Hazards. Inst. Chem Engrs. 1960 p 25.

13. Anthony E J, 'The use of venting formulae in the design and protection of building and industrial plant from damage by gas or vapour explosions'. J.Hazardous Mt'ls 2 (1977/78) 23-49.

14. Dragosavic M. 'Structural measures against natural gas explosions in high rise blocks of flats'. Heron 19 (4) (1973).

Figure 1   Explosion cell

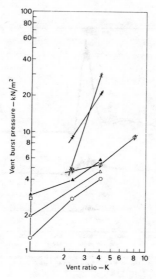

Figure 2 Vent burst pressure of various
vent covers used in the experiments;
For identification see Table 1

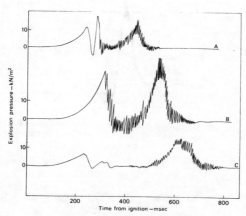

Figure 3 Typical explosion pressure traces with plastic vent covers, central
ignition. Trace A polyester (50µm) propane (4.3%), K = 2.3;
Trace B polyester (50µm) natural gas (10%), K = 4; Trace C
polyethylene (130µm) natural gas (10%), K = 4.

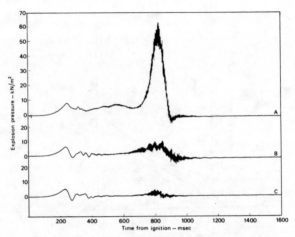

Figure 4 Effect of vent relief area on explosion pressure, fibreboard vents, natural gas (10%) central ignition. Trace A K = 8, Trace B K = 4, Trace C K = 2.4

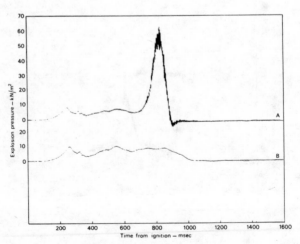

Figure 5 Effect of lining compartment with acoustically absorbent material (rock wool) fibreboard vents, natural gas (10%) central ignition. K = 8. Trace A no lining. Trace B with lining

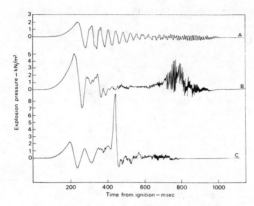

Figure 6 Effect of ignition location on explosion pressure, fibreboard vents (K = 2.3), natural gas (10%). Trace A ignition at centre of front face, Trace B ignition at geometric centre of compartment. Trace C ignition at centre of rear wall

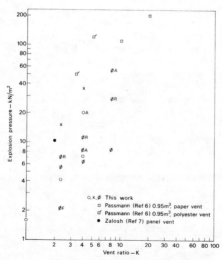

Figure 7 Explosion pressures for stoichiometric natural gas central ignition; Comparison of this work with published data. A indicates 'acoustic' vibratory peak corresponding to lower first peak shown below, R denotes ignition at centre of rear wall, F denotes ignition at centre of front wall.

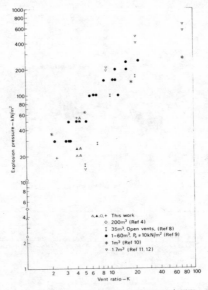

Figure 8 Explosion pressures for stoichiometric propane and pentane, central ignition, comparison of this work with published data. 'A' indicates an 'acoustic' vibratory peak

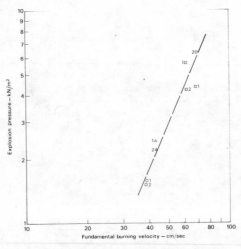

Figure 9 Relationship between maximum explosion pressure and burning velocity. Vent ratio K ≤1, vent covering polythene (130μm). 1 denotes first peak, 2 denotes secondary peak.

DUST EXPLOSION EXPERIMENTS IN A VENTED 500 m^3 SILO CELL

R.K. ECKHOFF and K. FUHRE
CHR. MICHELSEN INSTITUTE, BERGEN, NORWAY

NATURE OF THE VENTED DUST EXPLOSION

Basic features

The overpressure P(t) in a vented enclosure, in which a dust cloud deflagration takes place, will always be the net result of two simultaneous, competing processes:

- Burning of the dust, causing the pressure to increase.
- Flow of unburnt and burning dust cloud, and combustion products through the vent opening, causing the pressure to decrease.

When considering the two competing processes involved, it seems clear that predicting the rate of heat generation in the enclosure is by far the most demanding task, and a comprehensive theoretical treatment of this problem has not yet been undertaken[1]. Appreciation of the decisive influence of the heat generation rate on P_{max} attained in a given vented enclosure, created the motivation for performing the large-scale silo experiments to be reported here.

Factors influencing the heat generation rate

When trying to assess the instantaneous rate of heat production, or the accumulated released heat at any instant, during a dust explosion in a vented enclosure, several factors must be considered:

a) chemical composition of the dust, including moisture

b) distributions of particle sizes and shapes in dust, determining the specific surface of the dust in the fully dispersed state

c) degree of dust dispersion, or of agglomeration of dust particles, determining the effective specific surface available to the combustion process in the dust cloud in the actual industrial situation.

d) distribution of dust concentration in the actual cloud

e) distribution of initial turbulence in the actual cloud

f) possibility of generation of explosion-induced turbulence in the still unburnt part of the cloud

g) nature and location of the ignition source.

Factors a) and b) can be assessed accurately in laboratory tests provided representative samples are available, whereas factors c) to f) are determined entirely by the actual industrial dust cloud generation process, and the geometry of the confinement. The latter factors are very difficult to reproduce in laboratory experiments. However, laboratory tests have indeed demonstrated in general terms the importance of these factors[2].

The conclusion is clear: In future it will be necessary to discard the idea that the combustion rate in industrial dust clouds can be assumed to depend on the dust type only, i.e. solely on factors a) and b) above (K_{st}-concept). In addition,the practical circumstances under which the dust cloud is generated will have to be given careful consideration.

HOW CAN DUST CLOUDS BE GENERATED AND IGNITED IN LARGE SILO CELLS?

This question certainly has more than one answer. In case the main material to be stored in the silo is in itself sufficiently fine to give explosible clouds in air, explosible dust clouds are most likely to arise, at least transiently,somewhere in the silo whenever new material is charged into it, whether pneumatically or mechanically. If the main material is coarse, such as grain, explosible dust clouds may be generated by unburnt dust being blown into the silo by preceding explosions elsewhere in the plant. Dust could for example be injected through the various openings close to the silo top. Injection through the hopper exit at the bottom seems a more unlikely scenario. Another process of dust cloud generation could be that dust layers having accumulated on the inside of the silo wall and roof,are disturbed and dispersed into a cloud by air blasts or mechanical vibrations induced for example by preceding explosions elsewhere in the plant.

The identification of likely ignition sources is another central problem. Dust flames from preceding explosions entering the silo through openings at the top are one possibility. Dispersion of smouldering dust deposits is another. The possible roles of electrical and mechanical sparks are a topic of current dispute.

THE EXPERIMENTAL 500 m^3 SILO

Eight 500 m^3 steel silo cells at Boge, 45 km east of Bergen, were made available for the present investigation. Most of the experiments were conducted in one of these cells, but after its collapse due to excessive pressure development in the soya meal test, the instrumentation was transferred to a second cell,which was then used in the remaining maize starch experiments. A winding staircase extending right to the top of the cell in use permitted easy mounting and inspection of diagnostic instruments at various desired levels above the silo bottom. In order to enable experiments with various desired vent areas to be conducted,a strong steel grid was constructed across the entire top surface of the experimental silo cell, permitting any desired part of it to be blocked by bolting steel plates to the grid. A cross-section of the experimental silo cell with point of ignition and diagnostic instrumentation is illustrated in Figure 1.

The ignition source used was about 50 g of dried nitro-cellulose contained in a plastic bag and ignited by a pair of electrically fired Ce-Mg (100 J) fuse heads. The nitro-cellulose flame was fully developed approximately 0.1 s after firing the fuse heads, and it maintained its full size for approximately 1.0 s. The total energy liberated by complete combustion of the nitro-cellulose was about 200 kJ.

Four different types of measurements were performed during an explosion:

a) Eight specially constructed dust cloud extraction probes, distributed throughout the silo volume as indicated in Figure 1, were used in an attempt at measuring the dust concentration distribution in the silo just prior to ignition. These probes were used only in the wheat grain dust experiments, and they were only able to collect the finest particles in the dust cloud.

b) Three piezoelectric Kistler No. 412 pressure transducers were used for measuring the development with time of the explosion pressure inside the silo cell. In some tests only the top and bottom transducers were in use.

c) Three narrow-angle optical photo-diode probes were used for detecting the flame front arrival times inside the silo cell.

d) 16 mm, 25 frame/s cine recordings of the explosions were taken from a convenient position on the hillside about 100 m from the silo top.

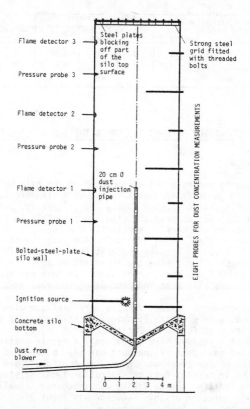

Figure 1. Cross-section of the experimental silo cells at Boge, Norway.

THE DUST USED IN THE 500 m^3 EXPLOSION EXPERIMENTS

Three different dusts were used. The first one was a wheat grain dust collected in the bag filters of the largest Norwegian grain silo. The second type was a soya meal supplied by another Norwegian company, whereas the third was a native maize starch quality obtained from Belgium.

A number of samples of the three dusts were taken from a representative number of bags on site at Boge and transferred to CMI in sealed containers for determination of particle size distributions, moisture contents, and P_{max} and $(dP/dt)_{max}$ in closed-bomb-tests. Samples were also sent to Fire Research Station, UK and Ciba-Geigy AG, Switzerland, for independent, parallel determination of P_{max} and $(dP/dt)_{max}$ in both the Hartmann bomb and the new 20 l sphere. The agreement between the results obtained in the various labs was quite satisfactory.

A representative summary of the results obtained is given in Table I.

TABLE I. Properties of dusts used in the 500 m^3 silo experiments.

Dust type	Median particle size by weight [μm]	Moisture content [weight %]	$(dP/dt)_{max}$ [bar/s]			P_{max} [bar(g)]		
			Hartmann bomb	20-litre sphere	H.b./20 l	Hartmann bomb	20-litre sphere	H.b./20 l
Wheat grain	20	12-15	120	325	0.37	6	7	0.86
Soya meal	50	9	25	175	0.14	4	6.5	0.62
Maize starch	15	10	300	425	0.71	7	7.5	0.93

GENERATION OF DUST CLOUDS IN THE 500 m^3 SILO CELL

The method used

It was decided to generate the dust clouds in the silo by simply blowing into it a known quantity of dust by means of a standard 120 Hp 'Vac-u-Vator' Model PTA 114144 (Dunbar Kapple Inc., Illinois, USA). As shown in Figure 1, the dust jet entered the silo approximately in the silo centre, with the direction of entry   vertically upwards. It is estimated that typical air velocities during dust injection into the silo were of the order of 10-15 m/s. To prevent the dust from escaping through the vent opening at the silo top during the injection period, the vent opening was always sealed with a thin sheet of plastic. In this way the dust cloud formation process was also kept independent of the vent area.

In all the experiments, except for the very last one with maize starch, the dust clouds were practically quiescent at the moment of ignition. This was achieved as follows: During the dust injection period the blower was operated at a steady, high speed until a fairly distinct change of its sound indicated that most of the dust had been blown into the silo. At this point the blower power was reduced to 'no load', and the automatic system for the sequential dust concentration sampling, start of Ampex tape recorder, and ignition source firing was triggered manually by the blower operator. The duration of this sequence was about 5 s. It should be noted that in the very last experiment with maize starch full dust injection was continued right through the entire explosion event.

Dust cloud structures

Although the same dust injection method was used for all the three dusts, the resulting dust clouds in the silo were probably quite dissimilar, because of differences in the dispersibilities and particle size distributions. In the case of the wheat grain dust, which is difficult to disperse because of high content of fibrous material, considerable quantities of unburnt dust (up to 100-150 kg) were normally found deposited on the silo bottom after the explosion. This clearly means that a significant fraction of the dust being blown into the silo cell, was in the form of large agglomerates, able to settle out of suspension before the dust cloud was ignited. However, this settling process apparently had a significant homogenizing effect on the concentration distribution of the remaining cloud of finely dispersed dust, inasmuch as the automatic dust concentration measurement system revealed a fairly even distribution of the fine dust fractions throughout the entire silo volume, both radially and axially.

Contrary to the wheat grain dust, the soya meal was very easy to disperse, and in this case the dust concentration distribution at the moment of ignition was probably fairly homogeneous throughout. However, no direct evidence of this is available because of the total collapse of the experimental silo in the soya meal experiment.

In the case of the maize starch, which is much finer than the soya meal, but still easy to disperse, there was probably a tendency of the dust concentration in the upper part of the silo being somewhat higher than in the lower part. This is because of the low settling rate in air of the fine maize starch grains (10-15 μm diameter). The delayed ignition that was observed in several of the maize starch experiments, is a further indication of the presence of such a concentration gradient. The dust concentration at the level of the ignition source was probably too low to propagate the flame at the moment of firing. However, the plastic bag containing the nitro-cellulose may have continued to burn for another few seconds, and long enough to ignite the edge of the settling cloud when its concentration close to the silo bottom had reached a level sufficient for flame propagation.

RESULTS OF EXPERIMENTS WITH INITIALLY QUIESCENT DUST CLOUDS

Comprehensive accounts of the work performed are given elsewhere[3,4]. The present contribution constitutes a brief summary, highlighting some of the important points.

About 330 kg of wheat grain dust (16 experiments), or 330 kg soya meal (one experiment only), or 150-200 kg of maize starch (6 experiments) was blown into the silo in each experiment. Hence, the nominal average dust concentration was about 660 g/m^3 both in the case of the wheat grain dust and in the case of the soya meal, and from 300-400 g/m^3 in the case of the maize starch. However, in view of what is said about the dust deposits at the silo bottom in the case of the wheat grain dust, the real average dust concentration in the silo volume was perhaps rather of the order of 400 g/m^3.

Explosion pressures

Apart from some characteristic 3-6 Hz oscillations of the order of 5-10 mbar peak-to-peak, and probably generated by standing waves in the dust-air-column in the silo, the pressure-time-histories were generally one single distinct bell-shaped peak of about 1 s duration. Two typical examples are given in Figure 2.

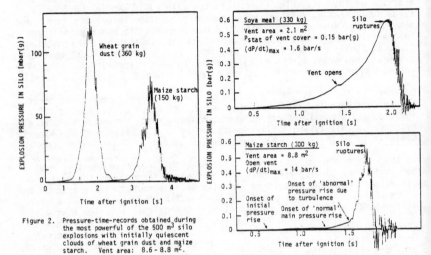

Figure 2. Pressure-time-records obtained during the most powerful of the 500 m³ silo explosions with initially quiescent clouds of wheat grain dust and maize starch. Vent area: 8.6 - 8.8 m².

Figure 4. Pressure-time-records obtained during the two explosions that caused the silos to rupture.

Figure 3 summarizes all the results obtained with quiescent dust clouds. Figure 4a gives the pressure-time-history of the soya meal explosion leading to silo rupture. It seems that the maximum pressure would probably not have been very much higher than 0.6 bar(g) even if the silo had been able to withstand the explosion. It is therefore suggested that the conservative envelope line in Figure 3 embracing all the CMI wheat grain dust data, is also applicable to both the maize starch and the soya meal containing at least 10% moisture.

Figure 3 also gives data obtained by Pinau et al.[5,6] (extrapolated from 100 m³) and by Matušek and Štroch[7] (500 m³ silo). It is not possible in the present context to discuss these two other large-scale investigations in relation to the present work. It should just be indicated that different dust cloud generation techniques were used, influencing both the dust concentration distribution, the degree of dust dispersion achieved, and not least the initial turbulence of the dust cloud at the moment of ignition.

For comparison Figure 4 also gives the vent area requirements specified by two codes in current use[8,9]. It should be emphasized that the ignition source used in the present investigation would indeed be classified as 'strong' (approx. 200 000 J).

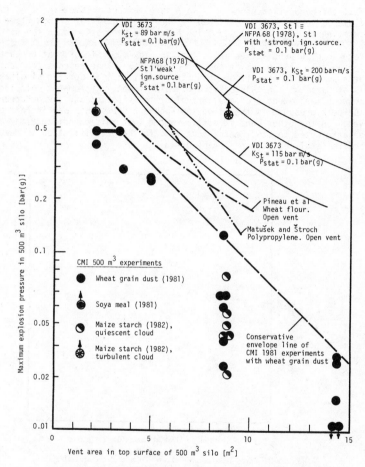

Figure 3. Maximum explosion pressures obtained in 500 m³ silo experiments with three different dust types. Results from two other large-scale test programmes. Vent area requirements by two existing venting codes.

Flame propagation

The vertical flame front speeds in the upper part of the silo were estimated on the basis of the flame-arrival-time-data provided by the three flame detectors. A typical trend for the wheat grain dust explosions would be a few m/s halfway down in the silo and subsequent acceleration to about 40-50 m/s close to the vent. In the case of the maize starch the few data available indicate a slightly higher speed.

It was observed that the duration of the flame signal at a given photo-diode station was considerably longer than the time required for the flame to travel from the bottom of the silo to its top. This suggests that the combustion process is to a large extent volumetric, i.e. the flame is very thick. This opposes idealized models of a thin flame sweeping through the cloud, which has often been assumed in the past.

Very long burn-out-times mean that the maximum rate of heat production in the silo will occur at the moment when the flame reaches the vent, because at this moment the quantity of dust that is burning simultaneously is at maximum. On this basis one would expect a systematic coincidence between the moment of flame arrival at the silo top and the occurrence of the peak pressure in the silo. As Figure 5 shows this was in fact observed. On the other hand it must be added that this coincidence may also be attributed to other mechanisms, and further clarification is needed before final conclusions can be drawn.

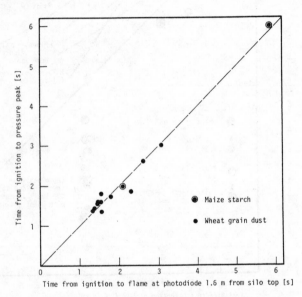

Figure 5. Correlation between data for time of flame arrival at silo top and time of peak of explosion pressure in silo.

RESULTS OF THE MAIZE STARCH EXPLOSION
INVOLVING TURBULENCE AND CAUSING SILO RUPTURE

The pressure-time-history

The pressure-time-history as given in Figure 4b consists of three phases. Following an initial small pressure rise of about 10 mbar after about 0.8 s, the onset of a slightly slower than 'normal' rise of pressure (≈ 0.17 bar/s) is observed at 1.3 s. The pressure continues to rise in this 'normal' manner up to about 55 mbar at about 1.5 s. The overpressure developed at this point is right in the middle of the range of peak pressures obtained in the 6 preceding 'normal' experiments (initially quiescent clouds). However, at this moment the 'abnormal', very fast pressure rise begins, resulting in a further increase of the pressure of more than 500 mbar in less than 0.2 s. The steepest part of this phase of the pressure-time-history has a slope of 14 bar/s, i.e. 80 times that in the 'normal' phase. The maximum overpressure recorded before the bin ruptured, was about 580 mbar. From the very high rate of pressure rise at the moment of rupture, it seems reasonable to deduce that had the bin been sufficiently strong the maximum pressure would probably have increased considerably more before reaching its peak.

Flame propagation

The average flame speed between 11.2 m and 6.2 m below the bin top was about 14 m/s. This velocity is of the same order as those obtained in the 'normal' experiments. The flame speed data thus provide a second indication, in addition to the pressure rise data, of the first phase of explosion development in this experiment being very similar to those in the preceding 6 shots where extensive turbulence was absent.

CONCLUSIONS

1. It must be expected that the properties of the dust cloud in terms of degree of dust dispersion, and distributions of dust concentration and turbulence, greatly influence P_{max} during a vented dust explosion.

2. In the present dust explosion experiments in a 500 m^3 silo cell with initially quiescent dust clouds, significantly smaller vent areas were required to maintain a given P_{max}, than those specified in the current VDI and NFPA codes.

3. However, it was also demonstrated that significant initial turbulence, in the form of a fairly strong turbulent dust jet, penetrating the upper part of the silo, increased P_{max} dramatically, by at least one order of magnitude, and even beyond the VDI code prediction.

4. A new, differentiated philosophy of dust explosion venting is needed, which must pay appropriate attention to the marked influence on the necessary vent area of the actual dust cloud generation and ignition processes. It is urgently needed to discuss and agree on, which dust cloud formation and ignition processes and associated combustion rates should, under various circumstances in industry, be regarded as the 'worst cases' that the vents to be designed should be able to master.

5. Further realistic large-scale experiments should indeed be encouraged.

ACKNOWLEDGEMENT

The authors are indebted to all the sponsors of both phases of the large-scale experiments undertaken, and to all persons and agencies providing other support and assistance. The invaluable assistance of M.H. Henery, S.J. Parker, G. Pedersen and H.G. Thorsen, all CMI, should be specially acknowledged. Sincere thanks are expressed to the population of Boge for their continued understanding and forbearance. Detailed further acknowledgements are given in references (3) and (4).

REFERENCES

1. Eckhoff, R.K.: 'Current dust explosion research at the CMI'. Proceedings of the International Conference on Fuel-Air Explosions held at McGill University, Montreal, Canada, November 4-6 (1981), edited by J.H.S. Lee and C.M. Guirao, University of Waterloo Press (1982), p. 657.

2. Eckhoff, R.K.: 'Relevance of using $(dP/dt)_{max}$ data from laboratory-scale tests for predicting explosion rates in practical industrial situations'. Paper to be given at the international VDI Colloquium 'Safe handling of flammable dusts' in Nürnberg 26-28. October (1983)

3. Eckhoff, R.K., 'Dust Explosion Experiments in a Vented 500 m^3 Fuhre, K., Silo Cell'. Henery, M.J., CMI-report No. 813307-1, June 1982. Parker, S.J., (Available from Chr. Michelsen Institute) Thorsen, H.G.:

4. Eckhoff, R.K., 'Maize Starch Explosion Experiments in a Vented Fuhre, K., 500 m^3 Storage Bin'. Henery, M.J., CMI-report No. 823307-1, October 1982. Pedersen, G., (Available from Chr. Michelsen Institute) Thorsen, H.G.:

5. Pineau, J., 'Efficacité des évents'. Giltaire, M., I.N.R.S.Note No. 1005-83-76 (38-07) Dangreaux, J.: (1976) CDU 614.839.

6. Pineau, J.: Private communication to R.K. Eckhoff in Paris (February 1982).

7. Matušek, Z., 'Problematik der Staubexplosionen und Massnahmen Štroch, V.: gegen Explosionsgefahren in Grossraumbunker für Schüttgut'. Staub-Reinhalt. Luft 40 (1980) No. 12, p. 503-510.

8. VDI-Richtlinie 3673 'Druckentlastung von Staubexplosionen' (1979).

9. NFPA 68 'Guide for Explosion Venting' (1978).

GASEOUS COMBUSTION VENTING - A SIMPLIFIED APPROACH

IAN SWIFT, FAUSKE & ASSOCIATES, INC., USA

INTRODUCTION

Although it is good industrial practice to avoid flammable atmospheres in industrial equipment and to remove all possible ignition sources, prudence, and many statutory codes, require that explosion protection be provided if the possibility of deflagration exists. A cost-effective, fail-safe solution is to provide sufficient venting to ensure the maximum pressure attained does not damage equipment. Lack of data spanning the conditions of interest to the chemical industry has hampered the development of universally accepted vent design methods - this is especially so for flammable mixtures initially at pressures greater than 1 atmosphere. The simple approach suggested here presents a method applicable to a wide range of initial conditions and avoids the use of "adjustable" parameters, for example to allow for the effects of turbulence.

THEORY

The complexity of the physical and chemical processes occurring during a deflagration - vented or otherwise - still defeats a completely theoretical treatment that would allow us, a priori, to predict flame movement. None of the current treatments [1,2,3,4,5,6,7] for closed vessel deflagrations includes flame-generated turbulence - which can enhance the rate of pressure rise - and all pertain to spherical vessels and one dimensional laminar burning. The analysis of Babkin et. al. [8] attempts to incorporate turbulence by redefining the familiar laminar burning velocity treatment in terms of a turbulent burning velocity - which must be determined by experiment. Unfortunately, the turbulent burning velocity is not an intrinsic property but is related to the spatially, and temporally, variable hydrodynamic state of the system. Flame-induced turbulence in closed volume deflagrations more generally occurs in large vessels and is a result of the amplification of flame front disturbances that can result from diffusional demixing, flame interactions with reflected pressure waves, convective lift, cellular combustion [9,10,11,12]. There is currently no satisfactory technique for predicting its onset or severity. The effects of creating turbulence (by fans, obstacles, or grids), namely an increase in the maximum deflagration pressure and the maximum rate of pressure rise, are well recorded [12,13,14,15]. If the situation with regard to closed vessel deflagration is unclear, that for vented deflagrations is best described as confused. Recent literature reviews on the subject confirm this [16,17, 18]. Bradley and Mitcheson's [16,17] work is the more complete since it also offers a sound theoretical basis for predicting vent areas; however, even the criteria they develop have been shown inadequate for large volumes with obstacle-generated turbulence [19]. An earlier unpublished report by Monsanto for the MCA Safety and Fire Protection Committee [20] reviewed available venting data and prediction methods and gave sample calculations. They concluded them to be insufficient to cover the range of real situations confronting the equipment and building designer. The most promising approaches to the problem are those that used the fundamentals of the combustion process (Munday [21], Yao et. al. [22], Pasman et. al. [23], Bradley and Mitcheson [16,17], and Crescitelli et. al. [24]). Techniques using similitude analysis [25], the method of characteristics [26,27], regression analysis [28-30], or nomograms [15] do little to advance understanding while playing a limited, but useful, role.

This approach is fundamental and avoids the use of adjustable "turbulence parameters" - it is similar in treatment to Bradley's work [4,16] - and expands on an earlier unpublished thesis [31]. The following assumptions are made:

1. spherical geometry gives the worst case conditions with central ignition;

2. the combustion gases are in chemical equilibrium;

3. ideal gas laws apply;

The physical situation is shown in Fig. 1, where a flame ball grows radially. At any instant the volumes of burnt and unburnt gases are given by

$$V_u = \frac{R_{Gu} T_u m_u}{P} \tag{1}$$

$$V_b = \frac{1}{P} \int_0^{m_b} R_{Gb} T_b \, dm_b \tag{2}$$

where V is the volume, R_G Universal Gas Constant, T is temperature, m is mass of gas, and b and u refer to burnt and unburnt respectively.

The unburnt gases will be adiabatically compressed

$$V_u = \frac{R_{Gu} m_u P_o^{\frac{1 - \gamma_u}{\gamma_u}} T_{uo}}{P^{\frac{1}{\gamma_u}}} \tag{3}$$

where γ is the ratio of specific heats, subscript o refers to initial conditions.

The mass of unburnt gas is given by

$$m_u = M - m_b - m_v \tag{4}$$

where, M is the total mass of gas initially in the sphere and m_v is the mass of gas lost through the vent.

Two flow regimes are possible for the flow of gas through the vent: if the ratio Pa/P is less than the critical pressure ratio P_c, the flow is sonic; if Pa/P > P_c, then the flow will be sub-sonic. P_c is given by

$$P_c = \left(\frac{2}{\gamma + 1} \right)^{\frac{\gamma}{\gamma - 1}} \tag{5}$$

For sonic flow the mass of vented gas is

$$m_v = CAB \int_0^t \left[\frac{P_0^{\frac{\gamma - 1}{\gamma}} \cdot P^{\frac{2\gamma - 1}{\gamma}}}{R_G \, T_0} \right]^{1/2} dt \tag{6}$$

$$B = \left(\left(\frac{2\gamma}{\gamma + 1} \right) \left(\frac{2}{\gamma + 1} \right)^{\frac{2}{\gamma - 1}} \right)^{1/2} \tag{7}$$

where C is the discharge coefficient for the vent, and A is the area of the vent.

For sub-sonic flow

$$m_v = CAD \int_0^t \frac{\dfrac{P_0^{\frac{\gamma - 1}{\gamma}} \cdot P^{\frac{2\gamma - 1}{\gamma}}}{R_G \, T_0} \cdot \left(\frac{P_a}{P} \right)^{\frac{2}{\gamma}} \left[1 - \left(\frac{P_a}{P} \right)^{\frac{\gamma - 1}{\gamma}} \right]}{\left[1 - \left(\frac{d_a}{d} \right)^4 \left(\frac{P_a}{P} \right)^{2/\gamma} \right]}$$

where P_a is ambient pressure, d_a is the vent diameter, d is the upstream diameter.

$$D = \left(\frac{2\gamma}{\gamma - 1} \right)^{1/2} \tag{9}$$

From volume conservation,

$$V = V_u + V_b + V_f \tag{10}$$

where V_f is the volume occupied by the flame front.

Substituting

$$V = \frac{R_{Gu} T_{uo} P_o}{P}^{\frac{1 - \gamma_u}{\gamma_u}} (M - m_b - m_v) + \frac{1}{P} \int_0^{m_b} R_{Gb} T_b \, dm_b$$

$$+ \frac{4}{3} \pi \left(r_b^3 - (r_b - \delta)^3 \right) \tag{11}$$

where r_b is the flame radius, and δ is the flame thickness.

Thus, solution of (11) gives the pressure for any value of the flame radius. However, the burned gas temperature is not uniform throughout the burnt gases, nor is the Gas Constant. After each element of gas is consumed, the pressure increases; this compression raises the temperature of both unburnt gases and previously burnt elements. This effect is shown diagrammatically in Fig. 2; the increase in the unburnt gas temperature means that the flame temperature of that element will be higher than the preceding ones. Although combustion reactions are thermodynamically irreversible, we can assume that the composition of the combustion products will follow changes in chemical equilibrium with the temperature. Thus, in calculating the burnt gas volume in (11), we must consider the history of each element, such that

$$\frac{1}{P} \int_0^{m_b} R_{Gb} T_b \, dm_b = \frac{1}{P} \int_{m_b - (1-f_n)\Delta m}^{m_b} R_{Gb} T_b \, dm_b \quad +$$

$$\tag{12}$$

$$\frac{1}{P} \sum R_{Gb1} T_{b1} f_1 \Delta m_1 + R_{Gb2} T_{b2} \left((1 - f_1) + f_2 \right) \Delta m_2$$

$$+ \ldots \ldots \ldots + R_{Gbn} T_{bn} \left((1 - f_{(n-1)}) + f_n \right) \Delta m_n$$

where f is the fraction of mass element Δm not in the flame front, and n refers to the nth mass element.

Introducing a time frame, the rate of consumption of the unburnt gas is given for a spherical geometry by

$$\frac{dm_u}{dt} = -4 \pi r^2 \rho_u S_u \tag{13}$$

where ρ_u is the unburnt gas density, and S_u is the laminar burning velocity.

Thus,

$$-dt = \frac{dm_u}{4\pi r^2 \rho_u S_u} = \frac{R_{Gu} T_u}{4\pi r^2 S_u P} \frac{dm_u}{} \quad (14)$$

However,

$$S_u = f(T_u, P) \quad (15)$$

The general form of this variation of burning velocity - derivable from thermal explosion theory - with temperature and pressure is given by

$$S_u = \left(\frac{T}{T_o}\right)^m \left(\frac{P_o}{P}\right)^n S_{uo} \quad (16)$$

where, for most hydrocarbons, $m = 2$, $n = 1/2$. By solving equations 11 and 14, the pressure at any time after ignition can be determined.

Thus far the effects of turbulence have been neglected; before including it in the model, a more detailed discussion of its nature and effects is needed.

The Effect of Turbulence on Burning Velocity

A detailed discussion of the nature and structure of turbulence is beyond the scope of this paper. It has been described as a three-dimensional, time-dependent motion in which vortex stretching causes velocity fluctuations to spread to all wavelengths between a minimum, determined by viscous forces, and a maximum, determined by the boundary conditions of the flow. Furthermore, it is strongly coupled with the energy release that occurs during combustion; this complicates an already complex problem.

It is not facetious to suggest that this subject is in a state of flux. Both small- and large-scale experimental data suggest that considerable accelerations are possible by promoting turbulence in confined and semi-confined fuel/oxidant mixtures. Since the review of Andrews et. al. [32], there has been progress in modeling flame-turbulence interactions. Ballal and others have investigated the structure of turbulence and propose a three-region model [33-40]. Others have also attempted to take into account the coherent eddy structure of turbulence [50,46]; further developments, based on probability density functions (pdfs), avoid the difficulties of calculating the flame shape and surface area by specifying the thermodynamic state using a progress variable and a time averaged mean velocity [41-45]. However, the formulation and closure of the pdfs are not known for most situations of interest, such that, although this approach has great intrinsic merit, no flame laws have yet been established to allow accurate predictions. Spaldings ESCIMO theory [47] of turbulent combustion attempts to combine the statistical nature of turbulence with detailed hydrodynamic and thermo-dynamic models of reacting flow. He uses a Langrangean approach for time dependent variables (concentration, temperature) and an Eulerian framework that utilizes time-averaged variables and correlations to described "sink" and "source" terms governing convective and diffusive transport. This is an attractive concept since it embodies all of the known physical processes occurring in reactive flows; however, a complete description of the mathematics is not yet worked out. The most useful current approach must rest with that of Abdel-Gayed et. al. [48], in which the ratio of turbulent to laminar burning velocity is obtained in terms of the turbulent Reynolds number and the ratio of the laminar burning velocity to the r.m.s. turbulent velocity. Reference [17] adds theoretical support to the correlations of a wide range of experimental data [49], which can be used as a predictive tool. The correlations of [49] are a series of graphs for different turbulent Reynolds number regimes; for practical use these were transformed into numerical form.

Two forms of turbulent interactions are dealt with - that generated by the flame before the vent opens and that created during the venting process. Although the interaction of acoustic waves with flames is a well-known and documented phenomenon [51,52], there is little to guide us in predicting the onset of the instabilities. Zalosh [29] phenomenologically links the flame oscillations to the fundamental resonant acoustic frequency - not unlike the earlier work of Smith and Kilham [53]. Shchelkin and Troshin [54] suggested a method for predicting the onset of autoturbulence (the term they use for flame-generated turbulence) that fitted experimental data.

They hypothesized that any freely expanding flame will contain irregularities in the flame zone (it will not be a sphere, but more like an orange); whether they expand and grow or die out will depend on the system properties. These wrinkles in the flame zone are thought to be caused by reflected weak shock waves (acoustic waves) and will smooth out if the characteristic time for the shockwave to traverse the burnt gas is less than the characteristic time for the flame front to travel one flame thickness. Expressed mathematically,

$$\frac{r}{c_s} < \frac{\delta}{S_u} \tag{17}$$

where r is the flame radius, c_s is the speed of sound in the burnt gases, and δ is the flame thickness.

Thus, the criterion for autoturbulence becomes

$$\frac{r}{c_s} > \frac{\delta}{S_u} \tag{18}$$

In order to calculate the increase in surface area due to wrinkling, it is assumed that, when criterion (18) was satisfied, the wrinkles could be viewed as cells (conical in shape) covering the surface of the flame ball and that initially the cells had a base radius equal to the flame thickness and a height equal to twice the cell radius. Thus, the total number of cells covering the flame ball will be the surface area of the sphere divided by the surface area of a single cell. As the flame continued to grow, the number of cells was kept constant, and the cell radius and height were recalculated to find a new total surface area. With autoturbulence the laminar burning velocity is unchanged; only the flame area is extended. When venting begins the situation changes. This is illustrated in Fig. 3. Initially, unburnt gases will leave the vessel; this induced flow will cause

- turbulence in the gases flowing towards the vent

- distortion of the flame ball from a sphere to an ellipsoid.

This is treated in the following way. A flow Reynolds number (Re) is calculated based on the flow velocity and the vent diameter; this is transformed into a turbulent Reynolds number (R_λ) using the relationship suggested by Abdel-Gayed and Bradley [49]

$$Re_\lambda = 5.927 \times 10^{-8} \, (Re)^{1.84} \tag{19}$$

The RMS flow velocity (needed in the correlation of turbulent burning velocity), u', was taken to be 5% of the actual maximum flow velocity in the vessel during venting. Thus, the instantaneous Re_λ and u' values are used to calculate the turbulent burning velocity.

The flame distortion, idealized to be an ellipsoid, will also increase the flame area - found simply by re-calculating the ellipsoid surface area having the same volume as the spherical flame. The major radius of the ellipsoid is taken to be the spherical radius plus half the elongation caused by the flow towards the vent. This elongation is calculated by equating the momentum of the gases leaving the vessel at any instant with that of those remaining, then, integrating with time to give the distance travelled by the mass hot gases. The eccentricity of the ellipsoid is taken to be the ratio of the spherical radius to the major elliptical radius. Unburnt gases are vented only until the flame is pulled to the vent; then the model assumes that only burnt gases will leave.

The instantaneous flame properties - adiabatic and isentropic flame temperature, species concentration, ratio of specific heats, sonic velocity, gas constant on a mass basis, density - are calculated in the usual way at each stage. This is simply achieved by equating initial and final enthalpies (or entropies) of reactants and products using JANAF [55] thermochemical data. The concentration of eleven major chemical species, CO, CO_2, H, H_2, H_2O, OH, N_2, NO, O and O_2 were obtained by evaluating equilibrium constants from values of Gibbs function.

APPLICATION

Chemical processes are frequently operated at elevated pressures, and protection against explosion under these conditions becomes a difficult design problem. The model was tested against data from tests at atmospheric and initially elevated pressures. Table 1 shows the correspondence between calculated maximum pressures and those obtained by Cousins and Cotton [56]. Table 2 shows a similar comparison for experimental data obtained by Union Carbide and Table 3 for data from Donat [57]. The model is conservative in that it overpredicts the maximum pressure. This is probably due to a combination of factors rather than a single model parameter, for example the choice of discharge coefficient, the actual rather than assumed vent area, the time taken for the vent to become fully open after it starts to rupture or become detached, the validity of the flame temperature calculations (heat losses are neglected), and non-spherical geometry. The discharge coefficient and vent opening time effects are examined using the model.

Table 1

Initial Pressure (MPa)	Vent Burst Pressure (MPa)	Vent Radio ($m^2/100\ m^3$)	Maximum Pressure (MPa)		Deviation %
			Experimental	Calculated	
0.203	0.246	25.33	0.377	0.398	5.6
0.203	0.901	25.33	0.970	1.115	14.9
0.203	1.163	25.33	1.287	1.349	4.8
0.203	1.370	9.91	1.439	1.776	23.4
0.405	0.543	9.91	1.659	1.783	7.5
0.405	1.756	9.91	2.500	2.796	11.8
0.405	2.583	9.91	3.134	3.458	10.3
0.405	0.543	4.86	2.252	2.410	7.0
0.405	1.962	4.86	2.983	3.437	15.2

Note: Experimental data from Cousins and Cotton [56] for 5% propane/air mixtures in a 3.2 10^{-2} m^3 vessel. Initial temperature 298 K, Su 0.45 msec^{-1}.

Table 2

Initial Pressure (MPa)	Vent Burst Pressure (MPa)	Vessel Volume (m^3)	Vent Diameter (m)	Maximum Pressure (MPa)		Deviation %
				Experimental	Calculated	
0.308	0.377	1.89	0.356	0.121	0.146	20.7
0.308	0.653	1.89	0.356	0.114	0.162	42.1
0.308	0.653	1.89	0.203	0.221	0.233	5.4
0.308	0.377	1.89	0.203	0.222	0.225	1.4
0.308	0.377	3.79	0.356	0.176	0.195	10.8

Note: Unpublished Union Carbide data for 10% methane/air mixtures initial temperature assumed to be 298 K, Su, 0.43 msec^{-1}.

Table 3

Initial Pressure (MPa)	Vent Burst Pressure (MPa)	Vessel Volume (m^3)	Vent Area (m^2)	Maximum Pressure (MPa)		Deviation %
				Experimental	Calculated	
0.10	0.15	1.0	0.20	0.20	0.221	9.5
0.10	0.15	1.0	0.10	0.25	0.385	35.1
0.10	0.15	10.0	1.20	0.20	0.260	23.1
0.10	0.15	10.0	0.75	0.25	0.337	25.8
0.10	0.15	10.0	0.55	0.30	0.398	24.6
0.10	0.15	30.0	1.90	0.20	0.321	37.7
0.10	0.15	30.0	1.30	0.25	0.388	35.6
0.10	0.15	30.0	1.00	0.30	0.437	31.4

Note: Experimental data from Donat [57] for propane/air mixtures. Initial temperature assumed to be 298 K, equivalence ratio 1.05 and S_u, 0.45 msec^{-1}.

Discharge Coefficient

For the calculations in Tables 1, 2, and 3, a constant discharge coefficient of 0.6 was used - valid for incompressible flow through a sharp-edged orifice. Although a coefficient closer to 0.8 would be a better reflection of real conditions for gas venting, 0.6 was retained to allow for uncertainty in the real vent area which can occur if vents do not open fully. Figure 4 illustrates the effect on the maximum pressure as the discharge coefficient varies from 0.6 to 1.0 for deflagrations in vessels of different volumes at the same initial conditions and vent opening pressure. The choice of discharge coefficient clearly allows considerable latitude in matching calculated results with experimental data.

Vent Opening Time

If the time taken for a vent to become fully opened after beginning to tear or dislodge is long (compared to the total deflagration time), the actual vent area will be significantly reduced for most of the venting sequence. The effect was modeled by making the area fraction available for venting a linear function of the ratio of time elapsed after the vent begins to open to the total vent opening time. Figure 5 illustrates this for vessels of different volumes at the same initial conditions and vent opening pressure. The effect is more pronounced in small vessels and illustrates the problems of extrapolating behaviors from small scale tests to large-scale operations.

Models such as this provide a design tool for predicting vent areas needed under industrial conditions or can be used to prepare graphs for easier application. Those shown in Figures 6a-6h were generated by predicting maximum explosion pressures in vessels of different volumes for different vent areas and various vent opening pressures. A discharge coefficient of 0.6 and a vent opening time of 1 msec were assumed. A propane/air mixture with an equivalence ratio of 1.05, initially at 298 K, was used, and the laminar burning velocity for this mixture was taken to be 0.45 msec^{-1}. Figures 6a-6h show that a constant vent area to volume ratio, as suggested by others [15], might not be conservative for larger volumes - where turbulence plays a greater role. Solberg's [58] work indicates a similar trend.

Conclusions

A simple combustion model has been described that embodies the effects of turbulence on the maximum deflagration pressure in a vented vessel. When tested against experimental data, the model appears conservative; this may reflect more the lack of information on important experimental details than its inherent correctness. However, it should provide a useful tool for vent design and in planning experiments to validate such designs.

REFERENCES

1. J. Nagy, J. W. Conn and H. C. Verakis, "Explosion Development in a Spherical Vessel," Bureau of Mines Report of Investigation, RI 7279, (1969).

2. H. E. Perlee, F. N. Fuller and C. H. Saul, "Constant-Volume Flame Propagation," Bureau of Mines Report of Investigation, RI 7839, (1974).

3. E. Kansa and H. E. Perlee, "Constant-Volume Flame Propagation: Finite-Sound-Speed Theory," Bureau of Mines Report of Investigation, RI 8163, (1976).

4. D. Bradley and A. Mitcheson, Comb. and Flame, 26, 201-217 (1976).

5. G. I. Sivashinsky, Israel J. of Technology, 12, 317-321, (1974).

6. A. M. Garforth and C. J. Rallis, Acta Astron., 3, 879-888, (1976).

7. V. S. Babkin and V. I. Babushok, Fizika Gorneya i Vzryva, 13 (1), 24-29, (1977).

8. V. S. Babkin, V. I. Babushok and V. A. Suyushev; Fizika Goreniya i Vzryva, 13 (3), 354-358, (1977).

9. P. F. Ivashenko and V. S. Rumyantstu, Fizika Goreniya i Vzryva, 14 (3), 83-87, (1978).

10. N. A. Strel'chuk, P. F. Ivashenko and V. S. Rumyantsov, Fizika Goreniya i Vzryva, 12 (5), 775-778, (1976).

11. V. N. Krivulin, E. A. Kudryavtsev, A. N. Baratsov, I. S. Glukhov, and V. L. Pavlova, Fizika Goreniya i Vzryva, 14 (6), 11-16 (1978).

12. J. Nagy, E. C. Seiler, J. W. Conn and H. C. Verakis, "Explosion Development in Closed Vessels," Bureau of Mines Report of Investigation, RI 7507 (1971).

13. G. F. P. Harris, Comb. and Flame, 11, 17-25, (1967).

14. N. Ono, M. Takushoku, M. Kurusu and I. Fukue, Nippon Kikai Gakkai Ronbun Shu, Part II, 41 (349), 2724-2732, (1975).

15. W. Bartknecht, "Explosions, Course Prevention Protection," Springer-Verlag, Berlin, Heidelberg, New York (1981).

16. D. Bradley and A. Mitcheson, Comb. and Flame, 32, 221-236, (1978).

17. D. Bradley and A. Mitcheson, Comb. and Flame, 32, 237-255, (1978).

18. E. J. Anthony, J. Hazard. Mat, 2, 23-49, (1977/78).

19. J. H. S. Lee and C. M. Guirao, "Factors that Influence Pressure Development in Closed and Vented Vessels," Paper presented at the 15th Annual AIChE Loss Prevention Symposium, August 18-19, 1981, Detroit.

20. W. W. Russell, "Monsanto Report on Explosion Venting," Monsanto Co., St. Louis, May 11, 1973.

21. G. Munday, "The Calculation of Venting Areas for Pressure Relief of Explosions in Vessels," Proceedings for the Second Symposium on Chemical Process Hazards, pp. 46-54, (1963), Institution of Chemical Engineers.

22. C. Yao, J. de Ris, S. N. Bajpai, and J. L. Buckley, "Evaluation of Protection from Explosion Overpressure in AEC Gloveboxes," Report for U.S. Atomic Energy Commission, FMRC Report Serial No. 16215.1 (1969).

23. H. J. Pasman, Th. M. Groothuizen and H. de Goojier, "Design of Pressure Relief Vents," C. H. Bushmann (Ed), Proceedings of the 1st International Loss Prevention Symposium, The Hague, Elsevier, Amsterdam, pp. 185-189, (1974).

24. S. Crescitelli, G. Russo and V. Tufano, J. Occ. Acc., 2, 125-133, (1979).

25. W. E. Baker, J. C. Hokanson, and J. J. Kulesz, "A Model Analysis for Vented Dust Explosions," Presented at the Third Internation Symposium on Loss Prevention and Safety Promotion in the Chemical Industry, September 16-19, (1980), Basle Switzerland.

26. M. A. Nettleton, Comb. and Flame, 24, 65, (1975).

27. M. A. Nettleton, Fire Prev. Sci. Technol., 14, 27, (1976).

28. W. E. Baker, E. D. Esparza and J. J. Kulesz, "Venting of Chemical Explosions and Reactions," Presented at the Second International Symposium on Loss Prevention and Safety Promotion in the Chemical Industry," September 6-9 (1977), Heidelberg, W. Germany.

29. R. G. Zalosh, "Gas Explosions in Roomlike Vented Enclosures," Presented at 13th Loss Prevention Symposium, 86th National Meeting of the AIChE, April 2-5, (1979) Houston, Texas.

30. J. Singh, "Sizing Vents for Gas Explosions," Chem. Enging. Sept., 103-109, (1979).

31. Ian Swift, "The Rate of Pressure Rise During an Explosion, Theoretical and Practical," M.Sc. Disertation, (1971), University of Leeds, Graduate Centre for Combustion and Explosion.

32. G. E. Andrews, D. Bradley and S. B. Lwakabamba. Comb. and Flame 24, 285-304, (1975).

33. D. R. Ballal and A. H. Lefebvre, Acta Astron. 1, 471-483, (1973).

34. D. R. Ballal and A. H. Lefebvre, Proc. R. Soc. Lond. A 344, 217-234, (1975).

35. D. R. Ballal, Acta. Astron. 5, 1095-1112, (1978).

36. D. R. Ballal, Proc. R. Soc. Lond. A 367, 353-380, (1979).

37. D. R. Ballal, Proc. R. Soc. Lond. A 367, 485-502, (1979).

38. D. R. Ballal, Proc. R. Soc. Lond. A 368, 267-282, (1979).

39. D. R. Ballal, Proc. R. Soc. Lond. A 368, 283-293, (1979).

40. D. R. Ballal, Proc. R. Soc. Lond. A 368, 295-304, (1979).

41. M. Champion, K. N. C. Bray and J. B. Moss, Acta Astron. 5, 1063, (1978).

42. P. A. Libby and K. N. C. Bray, Comb. and Flame 39, 33-41, (1980).

43. C. Dopazo, Acta Astron. 3, 853-878, (1976).

44. A. S. Monim and A. M. Yaglom, Stat. Fluid Mech., 2, 310-317, (1975).

45. V. L. Zimont, E. A. Meshcheryakov and V. A. Sabel'nikov, Fizika Goreniya i Vzryva, 14 (3), 55-62, (1978).

46. R. J. Tabaczynski, F. H. Trinker and B. A. S. Shannon, Comb. and Flame, 39, 111-121 (1980).

47. D. B. Spalding, J. Energy, 2 (1), 16-23, (1978).

48. R. G. Abdel-Gayed, D. Bradley and M. McMahon, Seventeenth Symposium (International) on Combustion, Combustion Institute, Pittsburgh, pp. 245-254, (1979).

49. R. G. Abdel-Gayed and D. Bradley, Sixteenth Symposium (International) on Combustion, The Combustion Institute, Pittsburgh, p. 1725, (1979).

50. J. Chomiak, Sixteenth Symposium (International) on Combustion, The Combustion Institute, Pittsburgh, pp. 1665-1672, (1976).

51. G. H. Markstein (editor), Nonsteady Flame Propagation, Pergammon, (1964).

52. G. H. Markstein, "Flames as Amplifiers of Fluid Mechanical Disturbances," Proceedings of the Sixth U.S. National Congress of Applied Mechanics, pp. 11-33, ASME, (1970).

53. T. J. B. Smith and J. K. Kilham, J. Acoust. Soc. Amer. 35, 715, (1963).

54. K. I. Shchelkin and Ya. K. Troshin, "Gas Dynamics of Combustion," Mono Publishers, Baltimore (1965).

55. JANAF "Thermochemical Tables," U.S. Dept. of Commerce, Clearinghouse for Federal, Scientific and Technical Information, PB 168370, (1965).

56. E. W. Cousins and P. E. Cotton, Chem. Enging, Aug., (1951), 133-137.

57. C. Donat, "Pressure Relief as Used in Explosion Protection," Loss Prevention, Vol. 11, p. 87-92, (1977).

58. D. M. Solberg, J. A. Pappas and E. Skramstad, Eighteenth Symposium (International) on Combustion, p. 1607-1614, (1981), The Combustion Institute.

Figure 1

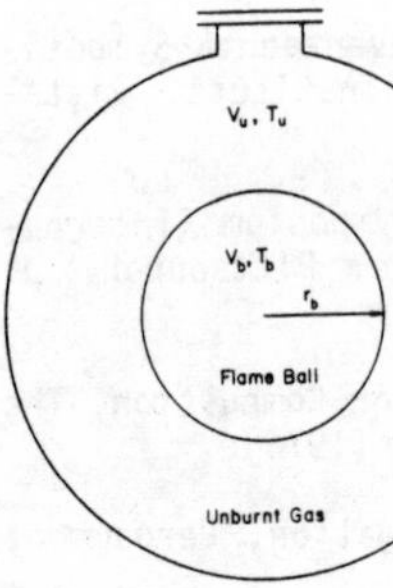

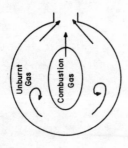

Flame Front Position Before Vent Opens

Figure 2

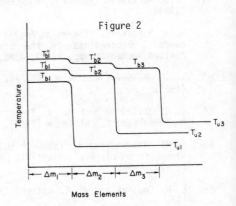

Mass Elements

Temperature Profile of Mass Elements
During Combustion and Compression

Figure 3

Gas Motion During Venting

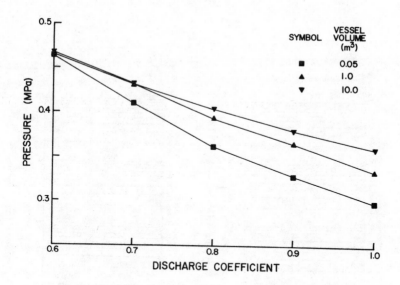

Figure 4   Effect of Discharge Coefficient on Maximum
Deflagration Pressure in Vessels of Different
Volume

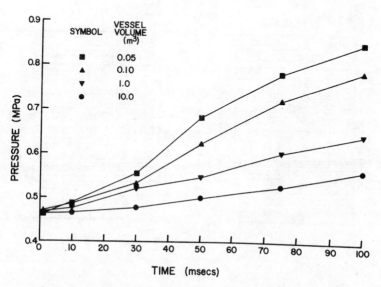

Figure 5   Effect of Vent Opening Time on Maximum Deflagration
Pressure in Vessels of Different Volume.

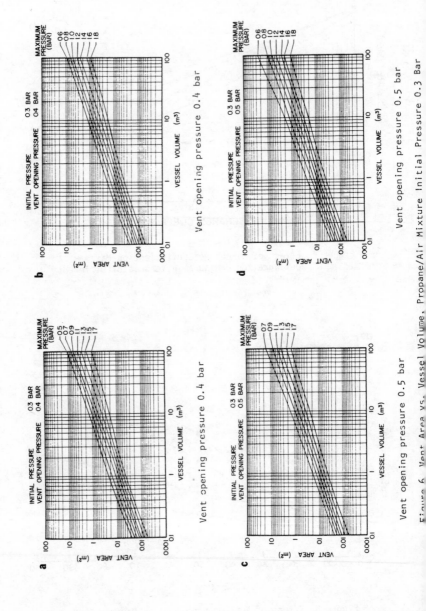

Figure 6. Vent Area vs. Vessel Volume, Propane/Air Mixture Initial Pressure 0.3 Bar

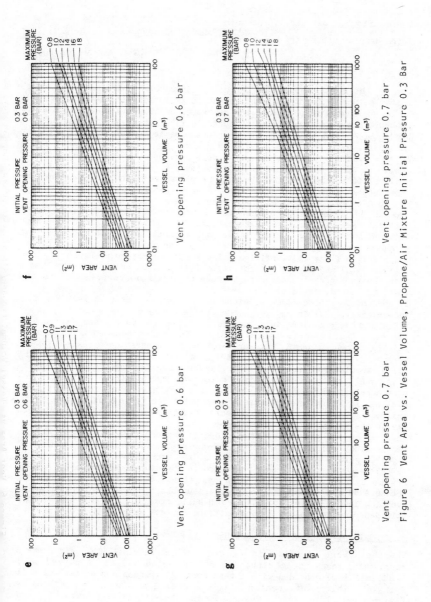

Vent opening pressure 0.6 bar

Vent opening pressure 0.6 bar

Vent opening pressure 0.7 bar

Vent opening pressure 0.7 bar

Figure 6 Vent Area vs. Vessel Volume, Propane/Air Mixture Initial Pressure 0.3 Bar

VENTING OF GAS EXPLOSIONS IN LARGE ROOMS

C.J.M.van Wingerden and J.P.Zeeuwen

Prins Maurits Laboratory TNO

1. INTRODUCTION

One of the most common ways to protect industrial buildings and process installations from internal explosions is explosion venting. The volume to be protected is provided with membrames or blow-out panels that allow unburned gases and combustion products to flow out. The vent opening must be large enough to limit the pressure rise to a value below the damage threshold.

In the past many correlations have been developed to select vent opening areas. Bradley and Mitcheson (1) reviewed these correlations showing a maximum difference of a factor of 6 between recommendations. Moreover experimental data on explosion venting were compared to a model developed by the authors. It appeared that discrepancies between model and experimental results occurred when the flame surface area was increased by external factors. Turbulence generation, for instance in a shear layer between out-flowing burned gases and unburned gases in the vessel as proposed by Solberg et al. (2), or flame instabilities such as acoustically driven flame instabilities (also known as Rayleigh-Taylor instabilities) (3) and Taylor instabilities (2) could be responsible for these flame surface area increases that cause the maximum overpressure in the model to be underestimate.

To account for these flame acceleration mechanisms other workers (such as Pasman et al. (4)) introduced a turbulence factor in their models. In (1) a "safe recommendation" is presented based on both models and experimental data (small-scale tests mainly). The safe recommendation for initially uncovered vents, however, appears to underestimate maximum overpressures found in a large room (35 m^3) by Solberg et al. (2). During their experiments strong flame acceleration mechanisms (turbulence generation and Taylor instabilities) occurred. Probably these mechanisms become more important when the volume of the vented room increases (see also (3)). Therefore in models accounting for flame acceleration mechanisms as in (4), a turbulence factor has to be adapted in a yet unknown way on the volume of the vented room.

In view of the above it was decided to perform large-scale experiments and to study the processes occurring during the venting in more detail in order to gain a better understanding of the mechanisms.

To this end experiments were performed in a medium-sized explosion vessel (capacity 5,2 m^3) provided with relatively large vents. The opening pressure of the vents was low in order to obtain reduced overpressures of such low levels as are acceptable for buildings.

This paper describes the experiments and surveys many of the results.

2. EXPERIMENTAL SET-UP AND PROCEDURE

The tests were performed in a 5,2-m^3 explosion vessel. The vessel has a rectangular shape (height and width 2 m, depth 1,2 m). A curved wall (radius 2,4 m) provided with a vent opening replaces one of the walls of 2 x 2 m^2. This vent opening (maximum area 2 m^2) could be reduced by fixing steel plates to the vessel. During these tests vent areas of 1 and 2 m^2 were used.

During most of the tests the vent was covered by two plastic sheets (thickness 0,1 mm) yielding a static response pressure of about 4,5 kPa. During some test two reinforced plastic sheets were used to obtain a higher vent opening pressure.

Gas mixtures were prepared by metering a calculated amount of combustible gas into the vessel. Mixing with air was effectuated by means of an electric fan that was switched off some time before the test so that the mixture was quiescent at ignition. A gas chromatograph was used to verify the composition of the mixture. Various concentrations of acetylene, ethene, propane and methane in air were investigated.

Usually the ignition source (an electric match head, ignition energy about 30 J) was located in the centre of the vessel. Some tests, however, were performed with the ignition source located near the rearwall or near the vent.

The pressure-time history of the explosion was recorded using two piezoresistive pressure transducers (Kulite semiconductor, type IPT-750-250) located in the rearwall and in one of the sidewalls of the vessel. Blast was measured outside the vessel at 4,25 m and 9,45 m from the vent and right in front of the vessel (Blast gauges: Celesco, type LC 33).

One of the sidewalls of the vessels was provided with a window through which the flame propagation could be recorded using a high-speed camera (Milliken). The vented explosions were filmed at a distance of about 25 m from the vessel using an 8-mm framing camera.

3. GENERAL OBSERVATIONS AND RESULTS

In general the pressure-time history shows a large number of fluctuations, three of which can be identified as "peaks" (see Fig.1).

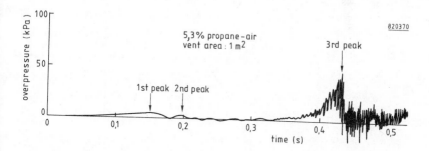

Figure 1. Example of pressure-time history measured in the 5,2-m^3 explosion vessel: Central ignition of quiescent mixture; static response pressure about 4,5 kPa.

The first peak is the result of the opening of the vent. After ignition the flame propagates as a sphere causing an internal pressure rise. When the bursting pressure of the cover is reached the cover bursts allowing unburned gases to escape. This results in a decrease in the internal pressure.

A large scatter in the effective (i.e. dynamic) opening pressures of the vent was found. The extent of this scatter is dependent first of all on how and where the plastic sheet first fails. In the second place it depends on the slowness of the opening of the vent in relation to the flame speed of the gas mixture. If for instance acetylene-air mixtures were used the first pressure peak amounted to maximal 8,8 kPa (2 m^2 vent area).

As the unburned gases flow out of the vessel, the expanding flame sphere is stretched towards the vent. Moreover, an analysis of the high speed films of the flame indicates that a Taylor instability develops in the region of the flame that is furthest away from the vent. These two effects cause the pressure to increase again after the first peak up to the moment when the flame reaches the vent as shown by the high-speed films. From this moment on combustion products of low density are vented instead of dense unburned gases. Owing to this the volumetric flow through the vent increases which results in the second pressure peak ($\underline{2}$).

When a 2-m^2 vent area is considered acetylene-air mixtures exhibit the highest second peak pressure (20,8 kPa). As acetylene and ethene were not investigated with a 1-m^2 vent area propane exhibits the highest second peak pressure (7,1 kPa) for 1-m^2 vent areas (see Fig.2).

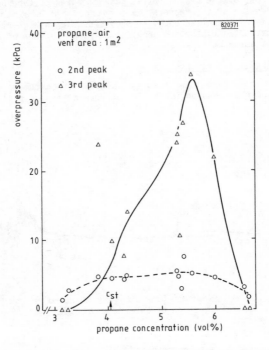

Figure 2. Amplitude of second and third peaks measured in vented propane-air explosions in the 5,2-m^3 vessel: Central ignition of quiescent mixture. Vent area: 1 m^2. Static response pressure about 4,5 kPa.

As a function of the composition of the mixture, the highest second peak pressure is found in mixtures that exhibit a cellular flame structure as a result of selective diffusion. In ($\underline{3}$) it is shown that these flames easily become unstable if exposed to a constant (Taylor instability) or periodic acceleration (acoustically driven instability). For propane and ethene the maximum second peak pressure is therefore found in fuel-rich mixtures and for methane in lean mixtures. Acetylene shows the maximum second peak in the fastest burning mixtures. Taylor instabilities seem to be of less importance for acetylene, possibly because they have not enough time to develop.

From the second peak on, the flow rate through the vent has increased to that only a very high burning rate is capable of generating a third peak. In the $5,2\text{-}m^3$ explosion vessel this third peak is caused by an acoustically driven flame instability, as can be demonstrated using the following observations:

a) The third peak under our experimental conditions is invariably accompanied with pressure fluctuations (see Fig.1) caused by a standing acoustic wave in the vessel. The frequency of the acoustic wave is about 200 Hz (for propane, ethene and methane) which under the circumstances in the vessel corresponds to the fundamental mode of the vessel when the wave stands between bottom and roof or between two sidewalls. For acetylene higher frequencies occur.

b) The third peak is very pronounced in mixtures which exhibit a cellular flame structure (Fig.2). As pointed out before these mixtures are very sensitive to periodic accelerations of the flame (acoustic waves). Acetylene shows a maximum third peak in the fastest burning mixtures.

c) If the driving force (the acoustic wave) is eliminated no third peak occurs, as was already shown in ($\underline{6}$). No other mechanism such as turbulence generated in shear layers seems to be strong enough to generate a third peak. The elimination of acoustic waves will be treated separately in section 6 of this paper.

The amplitude of the third peak (measured as the mean during oscillations) is comparable to that of the second peak in all cases, except for propane-air mixtures and a $1\text{-}m^2$ vent: in that case the third peak can be up to 34 kPa (see Fig.2), which is one order higher than the second peak. For methane-air mixtures and a $2\text{-}m^2$ vent no third peak can be distinguished.

If the ignition source is placed near the rearwall the third peak appears to be no higher than in the case of central ignition, and ignition near the vent results in very low third peaks (these tests were performed using different propane-air mixtures).

Some tests using propane-air mixtures and a $1\text{-}m^2$ vent area were performed with a cover consisting of two reinforced plastic sheets (static vent opening pressure about 12,0 kPa). As now the flame has already reached the vent when the cover is fully open the first two peaks coincide. The third peak, however, does not appear to be affected.

4. COMPARISON OF THE MEASURED MAXIMUM OVERPRESSURES TO A VENTING CRITERION

In Fig.3 the maximum overpressure for all tests where the second or third peak was higher than the vent opening pressure p_v is compared to the safe venting recommendation proposed by Bradley and Mitcheson ($\underline{1}$). Their criteria are based on the reduced parameter $\overline{A}/s_o$, where $\overline{A} = C_D \cdot A_v/A_s$ (C_D = vent discharge coefficient = 0,6 for sharp edged orifices, A_v = vent area and A_s is the total surfaces area of the vessel (15,8 m^2)) and

$\bar{s}_o = (\rho_u/\rho_b - 1)s_o/c_D$ (ρ_u and ρ_b = density of unburned and burned gased respectively, s_o = laminar burning velocity of the gas mixture and c_o = speed of sound).

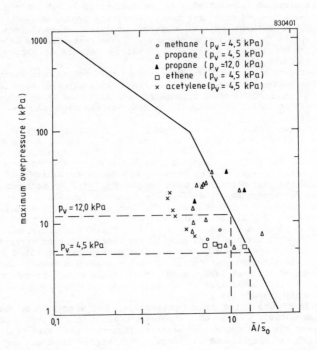

Figure 3. Maximum overpressure observed in vented explosions in the 5,2-m^3 vessel compared with the "safe recommendation" vent area for central ignition in near-spherical vessels proposed in (1).

The values of $\bar{s}_o$ of the gas mixtures used in the present study are based upon the data given in (1) and on data derived from flame speed measurements in a closed explosion vessel of 1-m^3 capacity.

The maximum overpressure in a vented explosion should not exceed the vent opening pressure p_v if $\bar{A}/\bar{s}_o$ is at least equal to the value prescribed by the solid line. As Fig.3 shows, this is not the case. Third peak overpressures measured for propane-air mixtures appear to be underestimated by the "safe recommendation".

5. EXTERNAL EFFECTS

Vented explosions can be a hazard to the immediate surroundings of the protected structure because of blast waves and burning jets coming from the vent. Few data on blast waves from vented explosions are available but recently some data were presented in (2) and (7).

An example of a blast wave measured during the present study is given in Fig.4. Every peak found in the vessel appears to be present in blast waves as well. Blast waves from the first peak appears to be very low (< 0,2 kPa at 4,25 m) if the vent opening pressure amounts to 4,5 kPa.

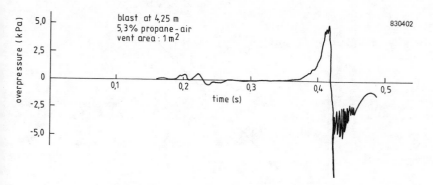

Figure 4. Example of a blast wave measured at 4,25 m from the vent of the 5,2-m^3 explosion vessel. Central ignition of quiescent mixture; static response pressure about 4,5 kPa.

If the opening pressure the vent is increased (12,0 kPa) the first peak in the blast wave becomes more distinct (0,8 - 1,0 kPa).
The blast wave properties of the second and third peak are shown in Figs.5a and 5b.
Fig.5a shows the relationship between the maximum internal pressure (1-m^2 vent area) and the maximum blast pressure at 4,25 m from the vent using both second and third peak results. As can be seen the blast pressure at 4,25 m is directly related to the maximum pressure within the vessel.
Fig.5b shows the relationship of the maximum pressures at 4,25 m and 9,45 m from the vent. It appears that blast decay with distance is in good agreement with Hopkinson's law ([8]) represented by the solid line. An exception to this is formed by the third peak from propane-air mixtures.

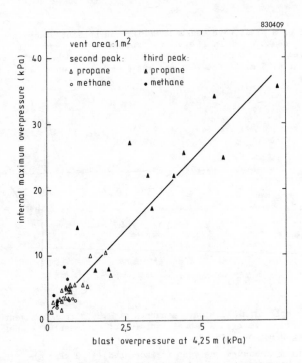

Figure 5a. Relationship between maximum overpressure measured in the $5,2$-m^3 explosion vessel and the maximum blast overpressure at $4,25$ m from the vent (vent area: 1 m^2).

Here blast decay seems to be stronger, which is possibly caused by the still reacting flame jet emerging from the vent during these tests. Similar results were found using a 2-m^2 vent. In this case, however, an anomaly occurs for the third peak of acetylene-air mixtures.

Qualitively similar results were obtained in ($\underline{7}$).

Especially during the third pressure peak strong burning jets emerging from the vent opening were found. Most of these jets exceeded a length of $5,5$ m measured from the vent.

As was also observed in ($\underline{2}$) the unburned gases vented to the surroundings after the opening of the vent (first peak $4,5$ kPa) were sometimes ignited by the flame emerging from the vessel after the second peak. This results in an external gas explosion. The behaviour of this gas explosion is quite different from the burning jets, because it expands almost spherically.

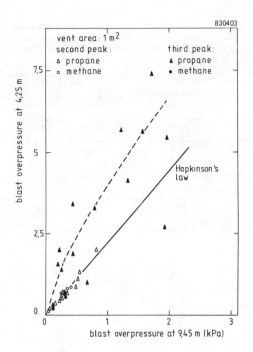

Figure 5b. Relationship between maximum blast overpressures at 4,25 m and at 9,45 m from the vent (vent area: 1 m^2).

6. REDUCTION OF THE MAXIMUM OVERPRESSURE BY LINING THE INTERIOR OF THE VESSEL

If occurrence of a third pressure peak could be prevented the maximum overpressure obtained during these tests would be reduced. As shown earlier the third peak is caused by the interaction of the flame and an acoustic wave. Therefore elimination of the acoustic wave should result in the elimination of the third peak.

To verify this the inner walls of the vessel were covered with different damping materials. First different walls were covered with glass wool (thickness 50 mm; results already presented in (6)). Later two different types of corrugated plates were applied (three walls covered).

Tests were performed using propane-air mixtures only. The results are shown in Table I. As the Table shows, the amplitude of the third peak is considerably influenced by applying different wall covers.

It should be noted that lining the interior wall does not influence other mechanisms such as turbulence generated in a shear layer.

TABLE I. Influence of lining the explosion vessel with damping materials on the pressure-time history of vented propane-air explosions in the $5,2-m^3$ vessel. Central ignition of quiescent mixture. Vent area 1 m^2. Static response pressure about 4,5 kPa.

Damping material	Lined vessel walls	propane concentration (%) v/v	amplitude (kPa)		
			1st peak	2nd peak	3rd peak
none	none	5,6	4,2	5,2	34,1
glass wool	sidewall + bottom	5,6	4,1	5,1	3,8
	2 sidewalls	5,5	4,1	5,7	3,1
	2 sidewalls + ceiling+bottom	5,3 6,0	3,6 4,0	5,8 3,4	0 0
corrugated plates [18cm ∏5cm]	2 sidewalls + ceiling	5,2	1,9	5,8	0
[8 cm ∏2,3cm 2,7cm]	2 sidewalls + ceilings	5,4	2,6	1,2	0

7. CONCLUSIONS

From the experiments the following conclusions can be drawn.
- When using low vent opening pressures the pressure-time histories exhibit three peaks. The first peak is caused by the opening of the vent.
  The second peak is governed by flame stretch and a Taylor instability,
  and occurs, when the transition to venting of burned gas has taken
  place. Especially for the smallest vent used in this study the third
  peak determines the maximum overpressure, which is caused by an acoustically driven flame instability. Damping of the acoustic oscillations
  by lining the interior of the vessel with different materials results
  in a substantial reduction in the maximum overpressure.
- Central ignition results in the highest maximum overpressures while an
  increase in the opening pressure of the vent from 4,5 kPa to 12,0 kPa
  has no significant influence on the maximum overpressure.
- Gas explosion venting models using turbulence factors based on small-
  -scale experiments underestimate effects in medium and large-scale experiments, as in the case when using a safe recommendation that is
  based mainly on small-scale tests.
  On the basis of current work and other studies it can be concluded that
  flame accelerating mechanisms such as turbulence and flame instabilities become more important with increasing volume of the room.
  Therefore maximum overpressures obtained during large-scale tests will
  be higher than is to be expected on the basis of small-scale tests.

More experiments in large rooms, preferably of different sizes and shapes are necessary to determine the role of flame accelerating mechanisms in these rooms and to see whether these mechanisms can be eliminated by adequate measures (as shown in this paper for the acoustically driven flame instability) or to arrive at appropriate venting criteria for cases in which a third pressure peak is to be expected.

ACKNOWLEDGEMENT

The authors wish to thank Mr.F.A.M.H.Jacobs for performing the experiments described in this paper.

REFERENCES

(1)   Bradley,D. and Mitcheson,A.: "The venting of gaseous explosions in spherical vessels. II - Theory and Experiment". Comb.Flame 32 (1978), p.237.
(2)   Solberg,D.M., Pappas,J.A. and Skramstad,E.: "Experimental investigations on partly confined gas explosions. Analysis of pressure loads, part I". Veritas report NV-79-0483, 1979.
(3)   Zalosh,R.G.: "Gas explosion tests in room-size vented enclosures". Paper presented at the 13th Loss Prevention Symposium, Houston, TX, 1979.
(4)   Pasman,H.J., Groothuizen,Th.M. and De Gooyer,H.: "Design of pressure relief vents", in: "Loss prevention and safety promotion in the process industries" (C.H.Buschman, ed.), Elsevier Scientific Publishing Co., 1974, p.185.
(5)   Markstein,G.H., in: "Non-steady flame propagation" (G.H.Markstein, ed.), Pergamon Press, 1964, p.93.
(6)   Van Wingerden,C.J.M. and Zeeuwen,J.P.: "On the role of acoustically driven flame instabilities in vented gas explosions and their elimination". To appear in Comb.Flame.
(7)   Palmer,K.A. and Tonkin,P.S.: "External pressures caused by venting gas explosions in a large chamber". Paper presented at the 3rd International Symposium on Loss Prevention and Safety Promotion in the Process Industries, Basle, Switzerland, 1980.
(8)   Baker,W.E.: "Explosions in air", University of Texas Press, 1973, p.54.

Special Design Problems

Chairman

R.C. Mill
Exxon Chemical Co., USA

PROTECTION OF DISTILLATION COLUMNS AGAINST VACUUM
CAUSED BY LOSS OF HEATING

J.S.Fitt (Consultant)*

> Loss of heat input to the reboiler of a distillation
> column can cause a rapid fall of pressure because the
> heat output to the condenser will be maintained until
> the vapour flow to it is reduced. With large distil-
> lation columns, the cost of designing for pressures
> much below atmospheric may be prohibitive. If the
> process materials are inflammable, the use of simple
> vacuum breakers may well be inadmissible. The al-
> ternative of inert gas injection requires estimation
> of the rate of pressure fall and the construction of
> a cooling curve.

> This problem is discussed with reference to a case
> study taken from the author's recent experience. Fol-
> lowing presentation of the example, the method used
> is discussed.

EXAMPLE: THE ETHYLENE DICHLORIDE 'HEADS COLUMN'

In this unit, low boilers are separated from the crude product from an
ethylene oxychlorination unit. On the plant in question, the normal
temperature and pressure of condensation were 336K and 1.14 bar, respectively. The design minimum pressure of the distillation column
was 0.9 bar; at the design minimum condensing temperature, 301 K, the
vapour pressure of the condensate is calculated to be 0.34 bar.

On an earlier unit, an atmospheric vent had been provided at the reflux
drum. The client arranged for an experiment to be carried out on the
older plant: steam flow to the reboiler was interrupted and it was
indeed observed that air flowed rapidly into the system, being (at
least partially) expelled a few minutes later. By the time the new
plant was planned, the public authorities had raised their standards
of environmental quality and prohibited the open vent as a means of
protection against vacuum. The decision was taken, at an early stage
in the design of the unit, to provide emergency injection of nitrogen
as an alternative safeguard. In the absence of an established method
for calculating the required injection rate, it was "guesstimated" as
100 m.³ per hour. When, at a later stage, a Hazard & Operability study
was performed, this problem was still unresolved. Therefore, a method
of calculation was then developed and applied - as described below.

---

* 1 Harvey Road, Guildford, Surrey GU1 3SG, England

The Problem.

To calculate the minimum rate of nitrogen injection required to ensure safety, we must:-

Estimate the heat capacity of the equipment and its contents

Derive a vapour pressure-temperature relationship for the overheads vapour

Calculate the rate of heat loss to the condenser as a function of still head pressure and temperature

Compute a vapour pressure-time curve for the first few minutes following the loss of steam to the reboiler

Use this curve to estimate the rate of nitrogen injection needed to prevent the pressure falling below the design limit

Repeat the calculation with allowance for the effect of added inert gas, to confirm that the correct rate has been chosen.

The Calculation.

Heat capacity. The estimation of this was of course straightforward. For lack of more accurate data, it was necessary to assume that specific heats of process fluids were constant over the temperature range considered.   The contents and metal masses of column, condenser and pipework were included.   In the first instance, the calculation was performed without allowance for metal heat capacity;   a correction for the latter was applied separately, as described below.

Vapour Pressure versus Temperature.   The composition of the normal overheads was:-

TABLE 1:

Component	Mol. Fraction
Ethyl Chloride	0.0527
trans-1-2-Dichloroethylene	0.0278
1-2-Dichloroethane	0.1503
Chloroform	0.5395
Carbon Tetrachloride	0.1819
1-1-Dichloroethane	0.0478

The vapour pressure of this mixture was found to be adequately represented by the equation:-

$$\ln p = 10.57 - \frac{3506}{T} \quad \text{(bar absolute)} \tag{1}$$

Heat Loss to Condenser.   In order to calculate the rates of condensation, two assumptions were made:-

Firstly, that the rate would be proportional to the ratio of vapour partial pressure to flowsheet vapour partial pressure

Secondly, that the rate would also be proportional to the condenser cold-end temperature differential.

By combining these assumptions, we get:-

$$\text{Rate of Condensation} = 0.1977 \times \frac{p}{1.14} \times \frac{T - 301}{35} \quad (\text{Kmol.s}^{-1}) \tag{2}$$

Vapour Pressure - Time Relationship.   As explained above, the initial assumption was that no heat would be abstracted from the metal masses; thus the heat loss to the condenser was supposed to be balanced by an equivalent fall in the temperature of the column contents.   The heat capacity calculation had shown that 2.112 Kmol. of liquid would have to be evaporated to produce a 1 K fall in the temperature of the system in the absence of an external heat source. Assuming that the rate of this evaporation would be equal to the rate of condensation calculated from equation (2), the time taken to produce that temperature reduction would be given by:-

$$\delta t = \frac{2.112 \times 1.14 \times 35}{0.1977p(T - 301)} = \frac{426.24}{p(T - 301)} \qquad (s.K^{-1}) \qquad (3)$$

This equation was used to calculate the time intervals required for the production of successive temperature falls of 1 K, from the initial value of 336 K down to 316 K. Since the proportion of non-condensibles is normally negligible, the corresponding vapour pressures may be taken as total system pressures.

The results of this calculation are shown in Fig. 1 (Curve I). It will be seen that an unsafe pressure would be reached, on these assumptions, after about 100 seconds from the time of loss of reboiler heat input.

Nitrogen Rate Required for Safety. It was assumed that the nitrogen valve would be actuated by a pressure trip set at 1.1 bar, that the valve would begin to open as soon as the signal was received and would take 15 seconds to open fully; it was considered that these assumptions were equivalent to supposing that the full nitrogen rate was available from the 25th second after reboiler failure. Since the vapour volume of the system was calculated to be 230m.^{3}, it was then possible to find the nitrogen rate required just to prevent the pressure falling below the design limit - by trial and error. The result is shown in Fig.1 (Curve II) and it will be seen that the requirement is for about 0.19m.^{3}s^{-1}(684 m.^{3} hr.$^{-1}$).

Allowance for Metal Heat Capacity. The simplest correction to apply is based on the assumption that the heat stored in the metal is immediately available. The effect of this is to increase all the times calculated by a factor of 1.742, in this case. The effect of this assumption is shown in Fig.2; the required nitrogen rate now becomes 0.11m.^{3}s^{-1}(396 m.^{3} hr.$^{-1}$).

The assumption of immediate availability of this additional heat is, however obviously unrealistic since a temperature difference must be produced to provide a driving force for heat transfer from equipment to process fluid. The calculation was repeated, therefore, assuming that heat recovery from the metal would only begin after the internal temperature had fallen by 5K. The results are plotted in Fig.3 and the refinement introduced will be seen to have increased the nitrogen requirement to 0.12m.^{3}s^{-1}(432m.^{3}hr.$^{-1}$).

These alternative corrections are compared in Table 2.

The Decision.

After much discussion, it was agreed that the lowest of these figures would be used. The extra refinement introduced by the temperature lag

TABLE 2:

Case	Time to reach design limit	Min. N$_2$rate for safety
	s	m.^{3}s^{-1}(m.^{3}hr.$^{-1}$)
1 Disregarding metal heat	100	0.19 (684)
2 Metal heat immediately available	170	0.11 (396)
3 Metal heat available after 5K δT	120	0.12 (432)

gives a correction of only 10% and this difference was felt to be of doubtful significance, given the inaccuracy of some of the data and also having regard to the many assumptions made in the calculation. Even this lowest figure is equivalent to 305 standard cubic metres per hour, over three times the original "guesstimate".

As a final back-up, an atmospheric vacuum breaker was added.

DISCUSSION

Apart from the special case of atmospheric and low pressure storage tanks (1), little or nothing has been published to guide designers on the protection of equipment against vacuum caused by rapid cooling. This is perhaps because equipment designed to operate at pressures below atmospheric is usually designed for full vacuum. Also, many of the cases in which protection would otherwise be required are ones where open vents would have been permitted in the days when standards of environmental protection were lower: improved environmental protection has lead to additional safety problems - as so often happens. The method of calculation used in the above example is a first attempt to provide a reasoned basis of design for the vacuum protection of a complicated distillation system, when the data available on physical properties were incomplete and the time available for producing a recommendation to the client was short - two days! To obtain a result many assumptions had to be made; the author is only too conscious of this fact but doubts whether a more rigorous approach, had it been possible, would have given materially different results.

The Assumptions.

Response of the System. A series of assumptions were made about the dynamic response of the system; they were considered to be reasonable because the time lapse of interest, after reboiler failure, was known to be short (from the client's plant trial referred to above) and because the control system was such that feed and reflux to the column would be expected to continue during the critical period. These assumptions were:-

- That the temperature, pressure and composition gradients would not change materially during the first few minutes after loss of reboiler heat input
- That pressure fluctuations would be transmitted instantaneously through the system
- That the rate of condensation would always equal the net rate of evaporation
- That the mean depth of liquid on distillation trays would not vary significantly.

Physical Properties. The ratio of specific to latent heat, for the process mixture, was assumed to be constant over the temperature range considered. It was assumed that the vapour pressure of the process liquid mixture could be represented by that of a pseudo-compound having the same boiling point. The latter assumption is justifiable because all the components (see Table 1) are chlorinated hydrocarbons with similar vapour pressure/temperature characteristics; the small amounts (altogether less than 0.6% molar) of water and inerts normally present were neglected.

Rate of Condensation. The assumption that this would be proportional to the partial pressure of condensibles and to the condenser cold-end temperature difference seemed intuitively reasonable. Use of the mean temperature difference would have introduced an unjustifiable complication.

Response of Nitrogen Injection Valve.   The assumptions made here are
critical because (Table 2) the time available is so short.   Even with
the use of quick acting valves, the set point for the pressure trip
must be quite close to normal operating pressure, with the attendant
risk of its being actuated un-necessarily. The combination of a trip
setting 0.04 bar below the flowsheet operating pressure and a valve
opening time of 15 seconds was consistent with the assumption that the
full nitrogen rate would be available at the 25th second after loss of
heat input.

Availability of Heat Stored in Metal Masses.   It was assumed that heat
from the metal of the column, condenser, reflux drum and pipework would
be available;   the metalwork of the reboiler was excluded on the
grounds that this was the most likely to lose heat to the surroundings,
particularly when the emergency were caused by loss of steam supply. The
assumption, in Case 3, that a 5K temperature difference would be required in order to release the heat from the metal into the process
fluid was equivalent to this extra source of heat becoming available
after about a minute.

The Chosen Solution to the Problem.

The choice of nitrogen injection as the means of protection against vacuum was preferred to the alternatives of interrupting the flow either
of vapour or of cooling water to the condenser.   These would have required large, expensive valves with a comparatively long response time;
the secondary consequences of such interruptions of flow would also have
been unacceptable as would those of erroneous actuation.   Fortunately,
the reliability of nitrogen supplies was extremely high;   on other
sites, the provision of emergency back-up might well have been necessary.

The work described in this paper was performed at a late stage in the
project, when the equipment was already in fabrication and the design
of pipework and control systems was already far advanced.   It was not,
therefore, possible to consider such expedients as mechanical strengthening of vessels or changing the configuration of pipework or the
design of venting arrangements.

The question remains whether the calculated nitrogen rates are not
either insufficient or else excessive, given the many assumptions which
had to be made in order to obtain a solution to problem in the time
available.   The results of the calculation do seem to be consistent
with the qualitative observations made during the plant trial on an older unit, referred to above.   Had quantitative plant experiments been
possible, they could obviously have provided a firmer basis for design;
in the absence of data from such experiments, few designers (and fewer
operators) would have been willing to dispense with the atmospheric
vacuum breaker - as a last line of defence.   High design rates are, of
course, in themselves an added safeguard but may strain the limits of
availability of supply, especially if there are a number of distillation
units to be protected.   In this context, rapid depressurisation would
be likely to cause frothing of liquid in downcomers, on trays etc. and
so the assumption that liquid depths would not vary may be excessively
pessimistic because such frothing would tend to damp down the initial
rapid fall in pressure.

RECOMMENDATIONS TO DESIGNERS

Ideally, the design problems of emergency vacuum protection should be resolved early enough for changes in design of equipment and control systems to be considered in order to minimise or eliminate the need for nitrogen injection. With smaller diameter vessels the cost of "full vacuum" design may well be justified. On the unit under discussion, pressure control was by a combination of throttling of a vent, coupled with by-passing of vapour around the condenser. If otherwise accep- table, the combination of an inert gas "bleed" with venting from the high point of the condenser to a small vent condenser, with this vent flow being throttled, should give rise to "inerts blanketing" of the main condenser at low pressures - with a selective spoiling of conden- ser efficiency. It would also have the advantage that the emergency system could be integrated with the normal pressure control to give a faster response.

Another advantage of tackling the problem early is that it would give time to obtain the data, possibly experimental, on which to base the calculation of rates by the above or by a more refined method.

If the designer is, however, in the usual position of having to make a preliminary guess at short notice, he should use the formula:-

$$F = 4(V_g - V_1)(1 - p_{min.}) \quad (m.^3 hr.^{-1}) \tag{4}$$

The result should be of the right order and may frighten his bosses into allowing him to do the job properly!

EQUATIONS (GENERAL FORMS) GIVEN IN MAIN TEXT

$$\ln p = A - \frac{B}{T} \qquad \text{(bar absolute)} \tag{1}$$

$$x = x_i \frac{p(T - T_f)}{p_i(T_i - T_f)} \qquad (\text{Kmol. s}^{-1}) \tag{2}$$

$$\frac{\delta t}{\delta T} = \frac{X p_i (T_i - T_f)}{x_i p (T - T_f)} \qquad (s.K^{-1}) \tag{3}$$

$$F = 4(V_g - V_1)(1 - p_{min.}) \quad (m.^3 hr.^{-1}) \tag{4}$$

SYMBOLS USED

A,B	= constants in vapour pressure/temperature equation	
F	= inert gas flow rate	$(m.^3 hr.^{-1})$
p	= vapour pressure	(bar absolute)
T	= temperature	(K)
δT	= temperature interval	(K)
t	= time	(s)
δt	= time interval	(s)
V	= volume	$(m.^3)$
X	= mass evaporated	$(Kmol. K^{-1})$
x	= mass condensed	$(Kmol. s^{-1})$

Subscripts.

i   = initial   f   = final

g   = gas phase   l   = liquid phase

        min.   = minimum

REFERENCE

1. Guide for venting atmospheric and low-pressure storage tanks;
   American Petroleum Institute; (1965).

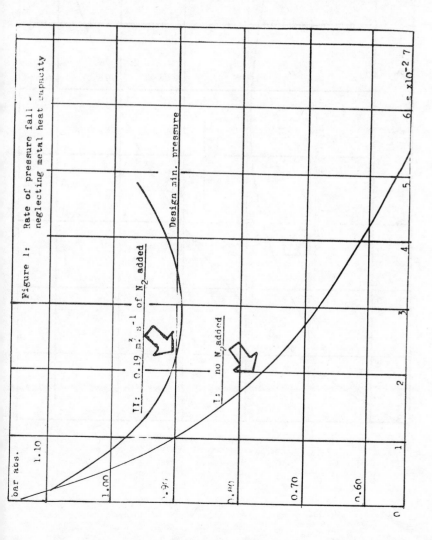

Figure 1:   Rate of pressure fall - neglecting metal heat capacity

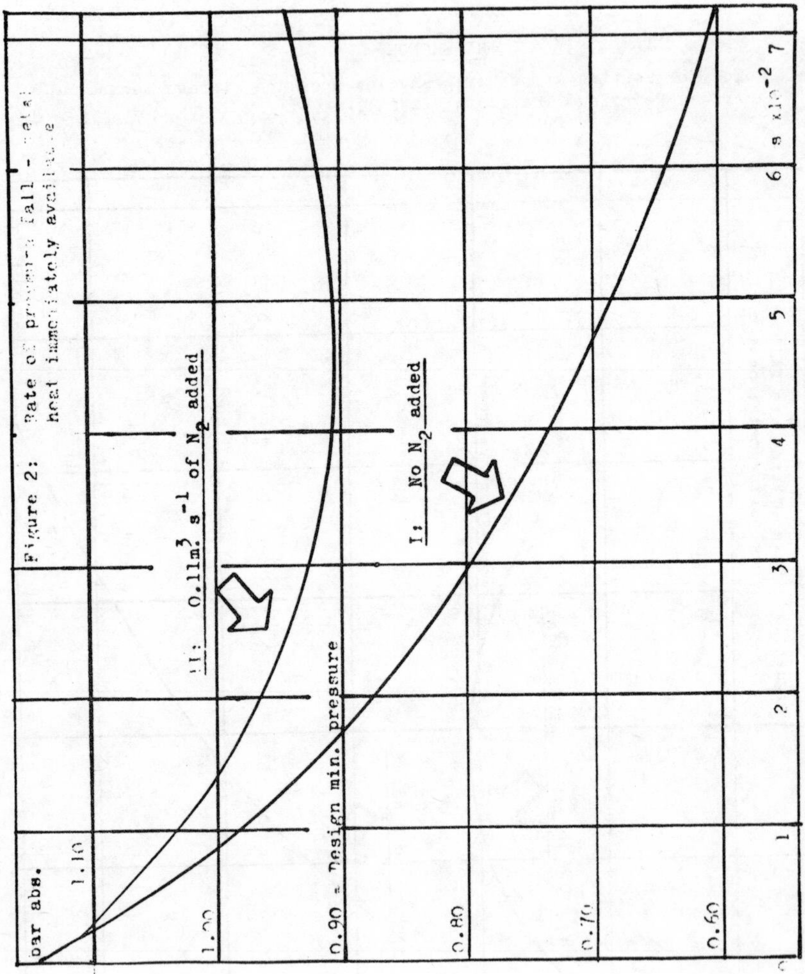
bar abs.
1.10
1.00
0.90 = Design min. pressure
0.80
0.70
0.60
II: 0.11m^3 s^{-1} of N_2 added
I: No N_2 added
0
1
2
3
4
5
6
7
s x10^{-2}

Figure 2: Rate of pressure fall - metal heat immediately available

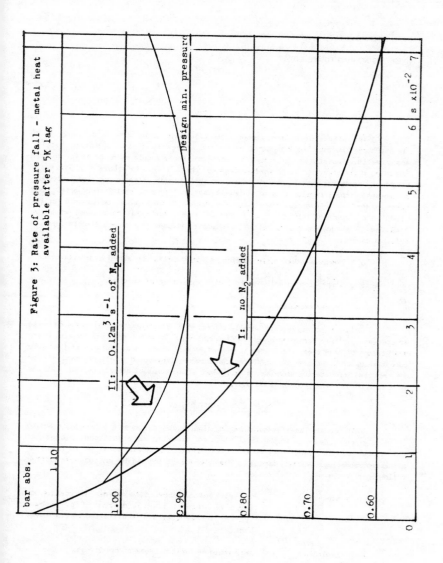

Figure 3: Rate of pressure fall – metal heat available after 5K lag

DETERIORATION OF STRUCTURES SUPPORTING REFINERY EQUIPMENT

B. N. Pritchard BSC, M.I.C.E. C. ENG.

(ESSO ENGINEERING EUROPE LTD).

INTRODUCTION

The safety of refinery operations often depends as much on the integrity of the structures supporting equipment as on the integrity of the equipment itself. Assuming the structures are properly designed, there should be no cause for concern regarding their integrity in the early stages of their life. Unfortunately, however, structures sometimes deteriorate with time and this deterioration is not always obvious. Consequently, the integrity of some structures may no longer be satisfactory after some time in service and worse, this fact may not be known to those responsible for the plant. The purpose of this paper is, therefore, to acquaint plant personnel with the problems of structural deterioration. The paper discusses the causes of deterioration and also covers:-

o   Measures taken during design and construction to minimise deterioration.

o   Inspection guidelines, including assessment of safety.

o   Repair techniques.

Refinery Equipment is usually supported by structures of reinforced concrete or steelwork, either bare or encased in fireproofing. The deterioration, inspection and maintenance of bare steel structures is generally well understood. This paper therefore concentrates on structures of reinforced concrete or fireproofed steelwork, considering in particular, deterioration due to corrosion. Structures of other materials, such as wood, are outside the scope of this paper.

CAUSES OF DETERIORATION

Old age in itself should have no significant effect on a structure, assuming it to be properly designed in the first place and assuming no significant change has taken place in its loading or environmental conditions. Nevertheless, we do find that structures sometimes exhibit damage. This damage is caused by one or more of the following effects:-

o   Overstress   —   Due either to accident, a change of loading or (rarely) improper initial design. Fig. 1 shows typical stress cracks in reinforced concrete members.

o   Explosion   —   This would be another cause of overstress.

o   Fire

o   Frost Damage   -   Either during the initial curing period or, for
                       instance, due to freezing of water trapped in bolt
                       pockets.

o   Chemical attack, including atmospheric corrosion - this is overwhelmingly
    the most common cause of damage.

Corrosion of Steelwork Structures

Atmospheric corrosion of bare steel structures is normally prevented by
painting. Painting techniques have been improved over recent decades and many years of
protection can normally be guaranteed. Providing they are designed so that all members
are accessible, inspection of painted steelwork structures is relatively simple, as is
their maintenance. When steelwork is hidden under fireproofing however, the onset and
degree of corrosion is not easily detectable. Properly designed and applied concrete
fireproofing protects the steelwork from corrosion. Nevertheless corrosion problems
sometimes occur and, as a result, paintingis recommended of all steelwork prior to
application of concrete fireproofing. Many of the various lightweight fireproofing
systems available offer no corrosion protection. Therefore, steelwork protected by
lightweight fireproofing, requires protection by a paint system, whose integrity at
critical positions should be regularly inspected.

Mechanism of Corrosion Protection by Concrete

Concrete protects buried steel in two ways:-

o   It provides a physical barrier to oxygen and moisture.

o   It passivates the steel.

In order to provide an efficient barrier to oxygen and moisture the concrete
should be dense and free of visible cracks. Since even dense, crack free, concrete is
not totally impermeable, the thickness of cover over the steel should also be adequate.

The process of passivation depends upon the alkalinity of the cement paste
binding the concrete. The corrosion rate of steel in tap water is far greater than when
it is in water containing dissolved cement. In tap water the corrosion potential of the
steel is quite high and iron ions are liberated to form $FeO(OH)$, a reddish dust. This
rust scale does not adhere to steel, so that it gives no protection against rapid
corrosion. In cement paste solution, the high pH results in a lower corrosion poten-
tial(called "Passivation") and a very adherent iron oxide, Fe_2O_3, is formed. This
oxide forms a continuous layer over the steel, protecting it from further corrosion.
When the pH at the concrete/steel interface is 11 or higher, corrosion is totally
inhibited.

Unfortunately, over the years, the alkalinity of the concrete is reduced, by
leaching and by carbonation effects. Both of these commence at the concrete surface and
become deeper as times goes by, the speed of penetration depending upon the permeability
of the concrete. Leaching is caused by the penetration of rainwater, steam or leaking
process water. Carbonation is caused by the combination of the alkaline calcium oxides
in the cement with carbon dioxide in rainwater to form calcium carbonate. Carbonation
also increases the permeability of the affected concrete. Given good concrete, a normal
environment and adequate cover to the steel, carbonation and leaching effects are not
normally a problem during the lifetime of a structure (1).

Effect of Atmospheric Corrosion

Many effects, including leaching, carbonation and attack by chemicals, can cause the breakdown of the protective oxide film on steel buried in concrete. Once this occurs, the action of oxygen ions which penetrate the permeable concrete, will cause surface corrosion of the steel. The corrosion products occupy about 8 times the volume of the uncorroded iron. The buildup of corrosion products between the steel and concrete therefore creates a high tensile stress, cracking and spalling the concrete. This reduces its effectiveness against replenishment of air and water, further accelerating corrosion of steel. In addition, the electrolyte is trapped by the concrete. Therefore corrosion is continuous, rather than intermittent, as in atmospheric corrosion of unprotected steel.

Effect of Fire on Concrete

When concrete is exposed to fire, it suffers in two ways:-

o   Mechanical damage is caused by high thermal stresses. This damage will take the form of cracking and spalling. Both effects reduce the physical protection to the underlying steel.

o   Carbonation is caused by the high temperature. The depth of this effect depends upon the surface temperature and the duration of its exposure to the fire. This effect often causes a pinkish tinge to freshly broken concrete.

As a result of both of these effects, the corrosion protection afforded to underlying steel by the concrete will be reduced. Following a fire it is therefore important to inspect the exposed concrete and repair any damaged material. Often it is advisable to paint the surface of concrete that has been exposed to fire, so as to supplement the corrosion protection to underlying steel. Guidance on suitable paints and other coatings is given in Ref. (2).

Effect of Chlorides

Chloride ions can readily penetrate the passive film covering the surface of the steel and initiate severe corrosion pitting. Once the film is broken down by chloride pitting, surface corrosion is initiated, with build-up of corrosion products leading to acceleration of the process in the way described above.

A threshold level of chloride ion concentration must be exceeded for oxide film breakdown. This threshold level reduces as the pH value decreases. The risk of corrosion is small, however, when chloride concentration is less than 0.4% of the cement content. When chloride concentration exceeds 1.0% the risk of corrosion is high, regardless of the pH value. (1) (3).

Sources of chloride ions include washing and spraying by saltwater (e.g. if firewater is salt), leakage of salt cooling water, atmospheric spray at seaside locations and calcium chloride added as an accelerator to the original concrete mix. Nowadays, the addition of calcium chloride during mixing of concrete is discouraged, but it was common practice some years ago.

Effect of Sulphates

Sulphates attack concrete very readily, reacting both with the free calcium hydroxide and hydrated calcium aluminate in cement. The products of these reactions occupy an increased volume and ultimately the associated bursting stresses break the bond between the aggregate and cement paste in the concrete. This shows up as a crazing and spalling of the surface.

The disintegration will weaken the concrete and also increase its permeability. Sulphate ions have a similar effect to that of chloride ions in causing a breakdown of the passive oxide film on the steel. Therefore accelerated corrosion will occur, as described above.

Sources of sulphate ions include sea water, some ground water and flue gases. This is, therefore, a problem in chimneys, sulphur plants. marine and some underground structures.

Sulphate attack of concrete is inhibited by modifying the content of the various constituents of the cement to form "sulphate resisting cement".

Effect of Acids

Being an alkali material, concrete is attacked by most acids. This attack removes the cement binder between the aggregate and causes disinte-gration. It is therefore important to provide surface protection, such as a coating of coaltar epoxy,to concrete expected to come into contact with acids.

Effect of Hydrocarbons

Concrete is very permeable to light hydrocarbons but it is not generally affected by them.

High Alumina Cement

In the past, high alumina cements have often been used, due to their rapid hardening, heat resistance and sulphate resistance. A number of disastrous failures however, have occurred in such structures. This has been due to conversion of the hardened cement paste to a weaker, softer material. Conversion is especially rapid when the atmosphere is warm and steamy, as is often the case in refineries. The conversion process also exposes any underlying steel to accelerated corrosion. Conversion can be kept within tolerable limits by adopting special mix, construction and design practices (4).

Cement/Aggregate Reaction (5)

This is a rare phenomenon which can occur when concrete contains certain siliceous aggregates, used with a high cement content and exposed to wet conditions. A gel is formed which disrupts the concrete. Aggregates should be chosen carefully to avoid this reaction.

PREVENTION OF CORROSION DAMAGE IN REINFORCED CONCRETE OR
STEELWORK PROTECTED BY CONCRETE FIREPROOFING

The best protection is given by providing adequate cover (30mm to 75mm, depending upon exposure conditions) of a dense concrete based upon an appropriate cement. The density of the concrete is improved by lowering the water/cement ratio during mixing and increasing the vibration applied during its placing. Addition of an air entrainment agent can improve the concrete impermeability. Proper curing is also important, so as to avoid the formation of shrinkage cracks during the setting process. National codes provide guidance upon cover, cement selection, concrete grades and curing procedures.

To prevent movement of moisture along the reinforcement or buried section, it is important to ensure good bond between steel and concrete. In reinforced concrete it is usually sufficient merely to clean off loose scale. In the case of steelwork sections under concrete fireproofing, however this does not appear to be sufficient and it is recommended that the steelwork be blast cleaned during fabrication immediately followed by two coats of lead chromate phenolic primer.

It has been found that cast-in-situ fireproofing concrete generally gives better protection than gunited concrete. The reasons for this are not entirely understood.

The shape of the concrete should be such that rainwater is easily shed. Alternatively, rainwater flashings can be used (see fig. 2). In conditions of exceptionally severe exposure, consideration should be given to the use of reinforcement coated with powdered epoxy or to coating the exterior surface of the concrete with a suitable material (2).

INSPECTION TECHNIQUES

After the onset of corrosion the concrete surface will crack along the lines of the steel. This is due to the bursting stresses accompanying the increase in volume of the corrosion products. Thus inspection at this stage is fairly simple, comprising examination of the external surface of the concrete for cracks, spalling and rust stains. Tapping the surface can also give valuable indications of laminar cracking, invisible externally.

Providing the damage is reported at an early stage in its development (while cracks are less than 0.5 mm wide), corrective measures are fairly simple. Once the damage progresses to the extent that it becomes readily apparent to a casual observer, it is often too late to avoid expensive repairs. The object of structural inspection should, therefore, be early detection of damage. This is best achieved by regular visual inspections by a qualified civil engineer. Such inspections should be carried out within 1 year of plant commissioning or of a significant change of duty or environment and at intervals of about 4 years thereafter. Formal reporting procedures should be adopted. This will facilitate the keeping of proper records, which can form the basis of a planned maintenance strategy.

During these inspections, the engineer should note all cracks, spalls and stains seen on the concrete. Cracks following the lines of reinforcement or the edges of underlying steelwork indicate possible corrosion. The position of reinforcement can be indicated by examination of the construction drawings, by use of a magnetic cover-meter or by a thermographic technique (6). If the location and direction of cracks is not related to underlying steel, they are probably not caused by corrosion and an in depth appraisal of the structure may be necessary to determine their cause. Stains are not always associated with corrosion of underlying steel, but are often caused by external sources, or sulphide inclusions in the concrete, neither of which is structurally significant.

Where cracking is parallel to the reinforcement or steelwork, a small section of the concrete cover should be carefully removed. Examination of the underlying steel will give an indication of the degree of corrosion. Spraying of phenolpthalein solution on the freshly broken surface will give an indication of the depth of carbonation. If the wetted surface turns pink (7), the concrete can be assumed to be sufficiently alkaline to protect the steel, provided chlorides or sulphates are not present in significant concentrations. Chloride and sulphate concentrations can be determined by analysis of samples of the concrete cover which was removed. A quick test is available (8), which gives a good indication of the chloride content of dust obtained by drilling 6mm diameter holes in the concrete cover.

Methods are available for detecting corrosion before external symptoms are apparent, but require specialist advice in their application and interpretation. The most promising of these tests is "Electrochemical Potential Mapping" (9). This involves measuring the difference of electrical potential between an exposed reinforcement bar or steelwork section and the surface of the concrete at various positions. The value of this potential difference gives guidance upon the state of corrosion in the underlying steel. Other tests involve chemical analysis of samples of the concrete. These may take the form of cores cut from the concrete. These cores are sliced into 25mm sections, each of which is analysed to determine its pH, the degree of carbonation and the concentration of chloride and sulphate ions. The analyses generally involve wet chemistry but scatter electron microscope examination of the polished surfaces of the sections have also been successfully used.

Gamma Radiography (10) and Ultrasonic Techniques (11) have also been used to detect corrosion and concrete cracking which cannot be seen from the surface.

The relatively sophisticated tests discussed above are not useful for general screening purposes. They should be limited to use in areas at particular risk (e.g. Fire damaged zones or structural elements in a severe environment) or as a follow-up investigation.

GUIDELINES FOR STRUCTURAL REPAIR

Following a detailed assessment of the extent and causes of the damage, a decision will be required on whether or not repairs are necessary and when they should be carried out. This decision will be influenced by:-

o   The cause of the damage.

o   The effect of the damage on the current safety of the structure.

o   The likely future rate of deterioration.

o   The importance of the plant supported by the structure and/or the
     hazards associated with structural failure.

o   The past performance and operating history of the plant.

o   Cost of repairs, versus the (maintenance) cost of doing nothing or
     the cost of complete replacement.

Generally, hairline cracks in concrete exposed to a refinery atmosphere can be
tolerated. If the cracks are a source of staining, however, it is probable that the
underlying steel is corroding and repairs should be considered regardless of crack
width. Also, some crack patterns, even though comprising fine cracks, indicate incipient
structural failure. It is therefore important that a civil engineer sees all cracks
reported in structural concrete.

Structural Safety Should Be Assessed

Assuming that the extent of the damage warrants repairs, an estimate should be
made of the safety of the structures. This estimate should be carried out by a qualified civil engineer. It should take account of the damage to the structure, and all of
the loads to which it will be subject.

In assessing the ultimate strength of the structure the characteristic
strengths of the materials should be used. When concrete has been subject to chemical
attack, this strength should be estimated using current test data, obtained by a
combination of Schmidt hammer, ultrasonics, bursting tests or crushing of cores. For
reinforced concrete, special attention should be paid to the effect on ultimate strength
of the loss of bond between reinforcement and concrete, due to corrosion of the reinforcement. Also, damage to horizontal ties in a reinforced concrete column can eliminate the contribution of the main vertical reinforcement to its compressive load
capacity.

The value of the safety factor, the rate at which deterioration is progressing
and the hazards associated with structural failure must be considered in answering the
following questions:-

o   Should the unit be shut down immediately?

o   Should repairs be carried out as soon as possible?

o   Can repairs be carried out while the unit is operating?

o   Can repairs be delayed until the next unit turnaround?

o   If so, are temporary repairs/structural reinforcement necessary?

o   Can a decision on repairs be delayed, monitoring damage meanwhile, in order to
     assess its rate of progress?.

Repairs Should Incorporate Measures To Remove Causes Of Deterioration

        Prior to initiating repairs, the causes of the deterioration should have been established. Appendix 1 provides guidance upon whether the causes of damage are still "active" (i.e. will continue to act after damage is repaired) or are "inactive". Wherever possible, the repair scheme should incorporate measures to eliminate or reduce the effect of "active" agents of deterioration. If this is not possible, it is likely that repairs will have to be repeated some time in the future. Typical examples of these measures are listed in Table 1 under the heading "Cures".

Repair Techniques

        Provided the underlying steel is undamaged and the concrete chemistry is acceptable, hairline cracks may be ignored and wider cracks may be sealed.Sealing is best accomplished by low viscosity epoxy, injected under pressure or applied using a vacuum technique (12). Where the cracking is extensive and is accompanied by spalling, it is usually necessary to cut back to sound concrete and replace with new concrete. Bond is improved by the application of a priming coat of polymer latex/cement mortar to the cut surface.

        If the underlying steel is corroded, or if the concrete chemistry is unacceptable at the steel/concrete interface, the concrete should be cut back. (Figure 3). In the case of reinforced concrete, the depth of cut should expose the whole circumference of the bar. The steel should be thoroughly cleaned and concrete replaced. The new concrete should comply with the recommendations previously listed, but addition of a polymer latex gives some advantage. While casting of the new concrete is preferred, it may sometimes be necessary to use a gunning technique. Where the thickness of the replacement concrete is so small its stability and bond are suspect, an epoxy should be used. In each case the boundaries of the cut should be normal to the concrete face, so as to avoid "feather edges".

        If unacceptable contamination by chlorides or sulphates is found at the outer surface, but acceptable conditions are found in the vicinity of the steel, action will depend upon circumstances. If the contaminants are likely to continue to attack the concrete, the affected concrete should be cut out and replaced and a protective coating applied. If the source of contaminant will be eliminated and ion concentrations in the concrete are fairly low, no action would be necessary. An alternative solution would be to apply a coating to the external surface which will inhibit penetration of the moisture and oxygen necessary for corrosion to take place. Many coatings are marketed, but only those should be considered with water vapour permeance less than 0.067g/SMN, high weathering and chemical resistance and with an ability to bridge any cracks that may be present.

Temporary additional support will be necessary during repairs if:-

o   The safety factor of the damaged structure is unsatisfactory, relative to the loads to be expected prior to completion of repairs.

o   The repair technique requires cutting out of concrete which may impair the safety of the structure (This includes the possibility of over-zealous cutting!).

CONCLUSION

In a typical refinery environment there is a good probability that conditions exist that could lead to deterioration of the structures supporting process equipment. This could have a serious effect on the integrity of the structure. In the case of structures of fireproofed steel and reinforced concrete, deterioration is not immediately obvious. Vigilence and an understanding of the causes and mechanism of deterioration is therefore necessary to detect the symptoms of deterioration at an early stage. This is best achieved by a system of regular inspections by a qualified civil Engineer coupled with a programme of planned maintenance.

REFERENCES

1. "The durability of steel in concrete; Part 2 - Diagnosis and assessment of cor-
   rosion-cracked concrete" - Building Research Establishment, U.K., Digest No. 264.

2. "The durability of steel in concrete; Part 3 - The repair of reinforced concrete" -
   building Research Establishment, U.K., Digest No. 265.

3. "International Seminar on Electrochemistry and Corrosion of Steel in Concrete - A
   State of the Art Summary" - Nigel Wilkins (Ed) - "Materials Performance" 1980 -
   National Association of Corrosion Engineers.

4. "Concerning the use of High Alumina Cement " - French Ministerial Circular 79 - 34,
   March 27, 1979.

5. "Alkali-aggregate reactions in concrete" - Building Research Establishment, U.K.,
   Digest No. 258.

6. "Location of reinforcement by thermography". B. Hillemeier, Hochtief A.G.,
   Frankfurt, West Germany.

7. "Carbonation of concrete made with dense natural aggregates" - M.H. Roberts,
   Building Research Establishment U.K., 1P 6/81.

8. "Simplified method for the detection and determination of chloride in hardened
   concrete", Building Research Station, U.K. IS 12/27.

9. "Half-cell potentials of reinforcing steel in concrete" - ASTM C-876-77. American
   Society of Testing Materials.

10. "Recommendations for non destructive methods of test for concrete" - Part 3 -
    "Gamma Radiography of Concrete" - BS 4408 Part 3 1970 - British Standards Instit-
    ution.

11. "Do, Part 5 - "Measurement of Velocity of Ultrasonic Pulses in Concrete, Appendix E
    - BS 4408 Part 5 1976 - British Standards Institution.

12. "Balvac speeds up repair jobs" - Contract Journal March 8, 1979.

13. "Deterioration, maintenance and repair of structures" - S. M. Johnson.

TABLE 1

COMMON CONCRETE DETERIORATION SYMPTOMS AND CURES [13]

The table presented below outlines symptoms of some common causes of concrete deterioration, its probable status and recommended cures. The table is mainly intended to aid refinery personnel in determining the cause of a particular deterioration problem and assessing future action.

Basic causes	Principal symptoms produced			Probable status of deteriorating agent	Cures
	Cracks	Spalling	Disintegration		
1. Occurrences during construction operations.	X			Inactive	seal cracks wider than hairline
2. Drying shrinkage	X			Inactive	
3. Temperature stress a. Variations in atmospheric temperature	X			Active	Reduce Thermal stress by:- o Removal of cause o Provide insulation o Add reinforcement
b. Variations in internal temperature	X	X		Active or inactive	
4. Absorption of moisture by the concrete	X	X	X	Active	Remove source or divert moisture or coat surface
5. Corrosion of the reinforcement a. Chemical	X	X		Active	Remove source of contamination. Cut out excessively contaminated concrete and replace.
b. Electrolytic	X	X		Active	
6. Chemical reactions	X	X	X	Active	
7. Weathering		X	X	Active	Coat surface
8. Explosions	X	X		Inactive	Seal cracks wider than hairline.
9. Erosion	X	X	X	Active	Remove cause. Provide anti erosion coating. Increase concrete cover.
10. Poor design details	X	X	X	Active	Correct/ Strengthen
11. Errors in design	X	X	X	Active	
12. Fire	X	X	X	Inactive	Replace cracked carbonated concrete. Coat surface.

FIG.1 TYPICAL STRESS CRACKS IN CONCRETE

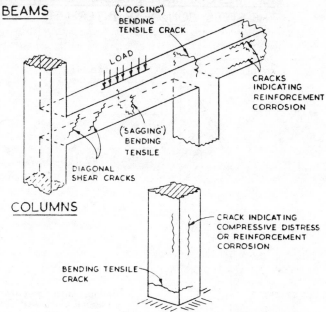

BEAMS

('HOGGING')
BENDING
TENSILE CRACK

LOAD

CRACKS
INDICATING
REINFORCEMENT
CORROSION

('SAGGING')
BENDING
TENSILE

DIAGONAL
SHEAR CRACKS

COLUMNS

CRACK INDICATING
COMPRESSIVE DISTRESS
OR REINFORCEMENT
CORROSION

BENDING TENSILE
CRACK

PLINTHS SUPPORTING DRUMS/EXCHANGERS

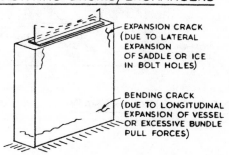

EXPANSION CRACK
(DUE TO LATERAL
EXPANSION
OF SADDLE OR ICE
IN BOLT HOLES)

BENDING CRACK
(DUE TO LONGITUDINAL
EXPANSION OF VESSEL
OR EXCESSIVE BUNDLE
PULL FORCES)

FIG 2 STEEL RAINCAPS OR FLASHINGS USED TO PREVENT WATER FROM PENETRATING THE STEEL CONCRETE INTERFACE

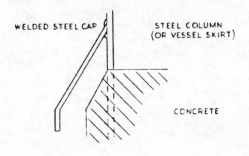

WELDED STEEL CAP

STEEL COLUMN (OR VESSEL SKIRT)

CONCRETE

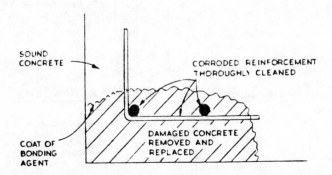

SOUND CONCRETE

CORRODED REINFORCEMENT THOROUGHLY CLEANED

COAT OF BONDING AGENT

DAMAGED CONCRETE REMOVED AND REPLACED

FIG 3 TYPICAL REPAIR

DETERMINING SAFETY ZONES FOR EXPOSURE TO FLARE RADIATION

G. FUMAROLA, D.M. DE FAVERI, R. PASTORINO, G. FERRAIOLO
Department of Chemical Engineering, University of Genova, Italy

SYNOPSIS

Among the combustion equipments for conversion of toxic or polluting substances into other less or non noxious, the flares show very peculiar problems from the engineering and the hygienic point of view.

Three kinds of impact with the environment are involved: noise, thermal radiation and air pollution.

As known thermal radiation generally plays a prevoiling role in the criteria followed for flare stack design.

An experimental research carried out in a wind tunnel has been dealing with observations of shape and length of the flame.

Based on the achieved results a predictive model has been developed for assessment of safety zones in terms of exposure to flame radiation.

FOREWORD

Safety zones around elevated flares for emergency releases are planned with respect to exposure levels to thermal radiation allowable for workers and structures.

Normally the technique used for this purpose is to treat the flame as a point source of thermal radiation, the location of the hypothetical release point being different from one model to another /1,2,3/.

This methodology follows clearly the side of caution leading to conservative predictions and wasteful overdesign.

With the object of avoiding this approximation other approaches have been proposed /4/, based on quite complex models of the flame, which involve too many parameters, with the result that the final equations can hardly be applied in practical instances.

An alternative model is here developed which simply considers the thermal radiation uniformly distributed along the axis of the flame.

The behaviour and the dimension of the flame have been investigated through scaled experiments in a wind tunnel /5/ and com

pared with two calculations models which are usually referred
to.

SHAPE AND LENGTH OF THE FLAME

Hoehne and Luce /6/ have studied the dispersion of turbulent
jets of hydrocarbons in a cross-wind through cold and hot flow
wind tunnel tests.

The results were used to derive a simple equation for the pre
diction of the shape of the flame axis

$$Z_t = 2.05 \; X^{0.28} \tag{1}$$

The end of the flame on the curve given by this equation corre
sponds to the point where the concentration of the flared gas
has decreased to the lean limit.

Experiments similar to those of Hoehne and Luce have been con
ducted in a wind tunnel on real flames from a scaled flare.

The results, given in Tab. 1, refer to two different fuels: me
thane and LPG, and to tests performed in the range of u_j/u_a from
4 to 76 with high Reynolds numbers.

Through the least square method the following equations have
been derived:

$$S_t = 3.8 \; C_t^{-0.48} \qquad \text{(for } C_t \leqslant 0.5) \tag{2}$$

$$S_t = 3.9 \; C_t^{-0.60} \qquad \text{(for } C_t > 0.5) \tag{3}$$

$$S_t = 3.2 \; X_t^{0.54} \tag{4}$$

$$Z_t = 3.1 \; X_t^{0.36} \tag{5}$$

Two equations (2 and 3) instead of a single have been derived,
analogously to Hoehne and Luce, in order to more properly sin
gle out the differentiation between the two regions within the
flame: jet-momentum dominated and wind dominated /5/.

The deviation between the two mentioned regions can be observed
in Fig. 1. The predictions given by equations (2) and (3) dif
fer a little from the corresponding by Hoehne and Luce.

Also eq. (5) differs from eq. (1) up to around 100%, at least
within distances from the source which are considered in plan
ning safety zones.

THERMAL RADIATION

Assuming the flame as a point source the thermal radiation in
a given point is usually calculated through

$$q = \frac{fQ}{4\pi D^2} \cos \theta \tag{6}$$

With the same approach, assuming a linear source uniformly distributed along the axis of the flame, one has

$$dq = \frac{fQ\cos\theta}{4\pi D_{x,z}^2} \, ds_t \tag{7}$$

The sight factor, $\cos\theta$, (see Fig. 2) can be expressed as

$$\cos\theta = \frac{z-\bar{z}}{D_{x,z}} \tag{8}$$

while $D_{x,z}$, on its turn, is

$$D_{x,z} = \sqrt{(z-\bar{z})^2 + (\bar{x}-x)^2} \tag{9}$$

Equations (4) and (5) can be rewritten as

$$z = 3.1x^{0.36}(d_jR)^{0.64} + h \tag{10}$$

$$s = 3.2x^{0.54}(d_jR)^{0.46} \tag{11}$$

substituting in eq. (7) one has

$$q = 0.043 \ fQ \int_0^{x_t} \frac{Ax^{0.36}+h-\bar{z}}{x\{(Ax^{0.36}+h-\bar{z})^2+(\bar{x}-x)^2\}^{3/2}} \, dx \tag{12}$$

where

$$A = 3.1(d_jR)^{0.64}$$

and

$$R = \frac{u_j}{u_a}(\frac{\rho_j}{\rho_a})^{1/2}$$

The calculation of the total thermal radiation in a given point, on the plane which contains the flame axis, can be derived through eq. (12) by numerical integration.

Eq. (12) has been compared with Brzustowski /1/ and API /2/ equations with reference to the working example specified in Tab. 2.

The results, given in Fig. 3, show similar gaussian profiles of the thermal radiation at the ground level for the three models, but different values of the maximum and of the distances downwind where they are located.

In particular, the maximum predicted by the model here developed is about 50% less than that by Brzustowski and 40% less than that by API.

A difference up to four times appears with regard to the loca‐
tion of the maximum.

NOMENCLATURE

$C = cu_j M_j / u_a M_a$

$X = x/Rd_j$

$Z = z/Rd_j$

$S = s/Rd_j$

D = distance from a given point, m

M = molecular weight, Kg/mole

Q = total heat release, Kcal/s

c = concentration of hydrocarbon gas, %

d_j = diameter of the flare stack, m

f = fraction of the radiant heat release

h = height of the flare stack, m

q = thermal radiation in a given point, Kcal/s m^2

s = distance along the axis of the flame, m

x = down-stream distance, m

z = cross-stream distance, m

x,z= coordinates of the given point, m

u = velocity, m/s

ρ = density, Kg/m^3

θ = sight angle

  subscripts

a= ambient air condition

j = discharge condition

t = end of the axis of the flame

REFERENCES

/1/ T.A. BRZUSTOWSKI, E.C. SOMMER - API Preprint 64, 1973.

/2/ Anon - API, RP 521, 1969.

/3/ G.R. KENT - Practical design of flare stacks.
    Hydrocarbon Processing, 43 (8) 121 (1964)

/4/ R. BECKER - Mathematical model of luminous flame radiation
    to determine safety zones and protective devices.
    3rd. Int. Symp. Loss Prevention and Safety Promotion in the
    Process Industries, Basle September 15-19, 1980.

/5/ G. FUMAROLA, D.M. DE FAVERI - Experimental approach to
    diffusion phenomena.
    Workshop on Advanced Mathematical Air Pollution Models, Col‐
    legio Ingegneri della Toscana, Firenze, March 19-20, 1979.

6. V.O. HOEHNE, R.G. LUCE - API Preprint 56-70 (1970).

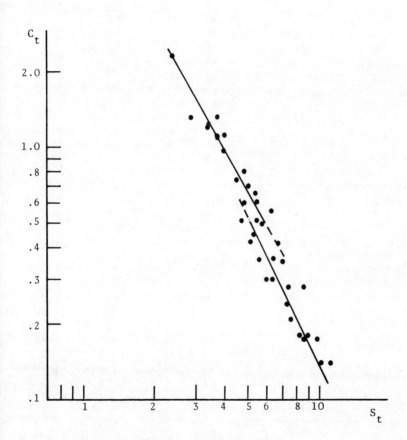

Fig. 1 - Experimental results referred in Tab. 1.

FUEL GAS	d_j (mm)	u_j/u_a	X_t	Z_t	S_t
		32	1.7	3.4	3.9
		37	1.5	3.4	3.7
		41	1.1	3.2	3.4
	1	41	1.3	3.1	3.4
		44	1.3	3.3	3.7
		44	0.9	2.7	2.9
		76	0.6	2.3	2.4
		10	3.2	5.0	6.2
		12	2.7	4.6	5.6
		12	3.7	4.9	6.4
METHANE	2	15	2.4	4.5	5.3
		17	2.1	4.0	4.7
		17	2.6	4.9	5.4
		20	2.1	4.4	4.8
		25	1.6	4.2	4.5
		6	4.9	6.4	8.1
	4	6	6.2	6.0	8.8
		8	4.8	5.2	7.3
		10	3.9	4.3	6.0
		16	3.6	4.8	6.3
	1	18	2.7	4.4	5.4
		23	2.7	3.9	4.8
		32	0.9	3.8	4.8
		8	5.6	6.5	8.6
		8	4.9	5.5	7.4
		10	4.1	5.6	6.9
LPG	2	12	3.1	3.7	5.1
		12	4.2	4.9	6.7
		14	3.4	4.4	5.8
		17	2.2	5.0	5.5
		20	2.2	4.7	5.1
		4	8.1	5.9	10.1
		4	7.7	7.7	11.1
	4	5	7.3	6.6	9.9
		5	6.4	5.5	8.5
		6	4.8	5.6	7.6

Tab. 1 - Experimental results achieved in a wind tunnel (Height of the scaled flare = 100 mm); $Re > 1.54 \cdot 10^4 (\rho_j/\rho_a)$.

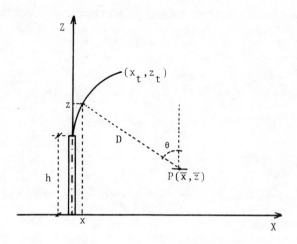

Fig. 2 - Diagram of the flare and the flame.

- Fuel gas	hydrocarbon vapors
- Flow rate	200,000 Kg/h
- Molecular weight of the fuel gas	46.1
- Discharge temperature	150°C
- Heat of combustion	11,000 Kcal/Kg
- Specific heat ratio	1.1
- Wind velocity	10 m/s
- Air temperature	27°C
- Fraction of the radiant heat release	0.3
- Low flammability limit	2 %
- Air relative humidity	50 %
- Vent exit velocity	0.2 Mach
- Height of the flare stack	50 m

Tab. 2 - Working example specifications.

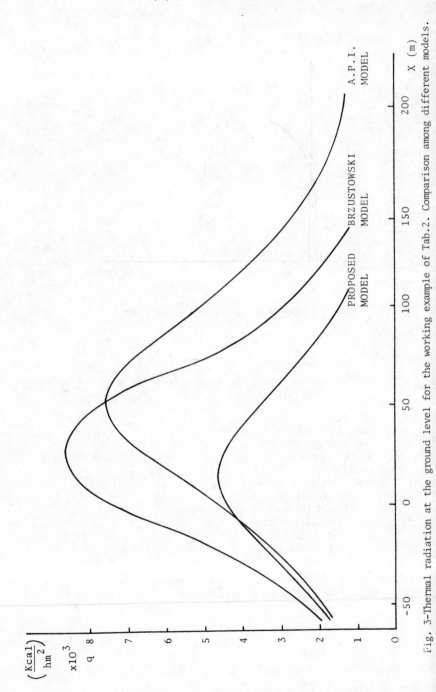

Fig. 3-Thermal radiation at the ground level for the working example of Tab.2. Comparison among different models.

Discussion

Reports of the discussions held on the papers
presented at the Symposium. Reports are
given for each Session and should be read in
conjunction with the papers in this volume.

Discussion

RUNAWAY REACTIONS

Rapporteur J. A. Hancock

Shell UK Oil

In the discussion of the first paper, 'Controlling Thermal Instabilities
in Chemical Processes', Dr. I. Swift (Fauske and Associates, USA) expressed
concern over the extent to which the "Test stand for evaluation of pressure
relief systems", described in Fig. 6 of the paper, could be used to
provide information for scale-up to actual relief conditions.

Dr. Schulz responded that, from tests simulating reaction fluids in
process without chemical reaction, useful data could be produced for the
assessment of hydraulic conditions at relief, by arranging artificial
head input to the reactor at such a rate as to produce in the test reactor
a vapour velocity (space velocity) equal to that expected in the full
scale reactor under runaway relief conditions.

Mr. H. S. Forrest (FMC, USA) cautioned that such data would be difficult
to interpret because, at equal velocities, there would be insufficient
time on the small scale for venting conditions to develop simulating
the full scale.

In response to a question from Mr. W. R. Klug (Wacker Chemie), Dr. Schulz
confirmed that methods of greater sensitivity are needed for the
measurement of exotherms where solid products are to be stored in larger
silos (ref. Fig. 4 of the paper) because heat dissipation, which is
largely by conduction, becomes more difficult as silo size increases.

Regarding the second paper 'Runaway Reactions in a Polyacrylonitrile Wet
Spinning Plant', Dr. O. Klais (Hoechst AG) was sceptical over the extent
to which the kinetic model, determined for a four step chemical reaction
from an experiment carried out in a single stage using a Sikkarex
apparatus supplemented by product analyses, could be extrapolated with
safety direct to the operation of a full-scale production plant.

Mr. Tufano was confident that it could in the case described.

In the discussion on the paper 'Thermohydraulic Processes in Pressure
Vessel and Discharge Line During Emergency Relieving', Dr. Friedel
emphasised that all the work described in the paper was carried out in
the absence of chemical reaction.

Dr. I. Swift (Fauske and Associates, USA) commented that the conclusion
of the paper, regarding the advisability of relieving a vessel with
first a small vent and then a large one, in order to minimise the extent
of two phase flow, while it might produce an optimum result in the case of
a vessel without chemical reaction would not necessarily do so where
chemical reaction runaway was the cause of the release.

Dr. Friedel agreed.

Mr. J. S. Binns (Diamond Shamrock) suggested that where a limited initial relief rate was preferred, for the reasons given in the paper, relief might better be achieved by proportional opening relief valves than by bursting discs.

Dr. Friedel accepted this.

In response to other comments, Dr. Friedel indicated that his group was hoping to extend the use of the test facility described in the paper to cover:-

- trials with viscous liquids
- trials with liquids already boiling at the point of opening of the vent
- trials with foaming liquids

Mr. A. R. Clare (Beecham Pharmaceuticals) suggested that work might also be useful on liquids containing solids in suspension.

Mr. A. P. Cox (Shell International) thought that temperature measurements within the pressure vessel of the test facility would have helped with the interpretation of the results obtained.

UNSTABLE PRODUCTS

Rapporteur Dr. D.A. Lihou

HELPS Ltd.

Self-Accelerating Decomposition of Ethylene Cyanide

In response to various questions of clarification, Dr. Löffler made the following comments: With adiabatic storage tests and isothermal DTA the induction time versus temperature was the same, nor did they find any effect of sample size in the range 100-200 gm. Therefore, he did not expect that the dependence of induction time on temperature, as shown by Figure 3, to be different for larger masses of ethylene cyanide. The activation energy derived from Figure 3 is 80 kJ per mole.

Isothermal DTA is the best method to screen compounds for self-accelerating decomposition characteristics. The experimental results are adequately described by the mechanism shown on page C3 and the authors did not believe that a detailed kinetic/thermodynamic model was necessary for ethylene cyanide; their recommendation is that prolonged isothermal DTA tests should be used more often when screening for thermal instability.

Thermal Explosion of Liquids

There was considerable interest in the physical phenomena occurring during deflagration of the foam: The deflagration takes place at the interface between the bubbles and the liquid; it is due to the compressibility of the two phase mixture which allows a much higher sonic velocity than in either the liquid or the gas. Dr. Verhoeff is convinced that condensed phase explosions in liquids cannot occur in the absence of gas bubbles. A film of an unstirred column of liquid showed that you can get a high intensity thermal explosion at low heating rates. The deflagration is unstable so that on a larger scale one could get detonation. The gas dynamics which permit transition to detonation need further study. The three stages of thermal runaway followed by initiation and finally explosion occurs whenever liquids explode; but not all thermal runaways produce explosions.

More than one questioner referred to the temperature differences between the top and bottom of the liquid. The upper temperature is higher because the concentration of bubbles near the free surface is highest and further decomposition of the gaseous decomposition products is occurring more rapidly than the primary decomposition in the liquid. Calibration of the test cell with an inert liquid showed that the thermal capacities of the liquid and of the inner vessel were similar.

Dr. Verhoeff said that he had no experimental results to show the effect of pressure on explosion sensitivity; nor the effect of viscosity and scale on the power of the explosion. They had not measured gas generation rate; but that was possible.

Experimental Principles for Predicting Safe Conditions for the Storage of Bulk Chemicals

Dr. Gygax agreed that the method he presented for determining critical
storage temperatures applies only when there is no autocatalysis. But
the gathered experimental data would provide evidence of this if it
occurred.   The various standard calorimetric tests are not suitable for
powdered solids which cannot be stirred and which have low thermal
conductivity.   They would not provide data on the rate of release of
heat of reaction.   Dr. Gygax considered that the assay method he used,
Figures 6 and 7, required less extrapolation to obtain the critical
temperature.   He admitted that a storage temperature of $30^{\circ}C$ would be
difficult to achieve for 100 l. drums in distribution; but he pointed
out that the actual critical temperature for drums of this size was $50^{\circ}C$.
The reason being that the low thermal conductivity, which made self-
heating so critical, was a benefit in the rise of temperature at
locations far from the drum wall.

Pressure Increase in Exothermic Decomposition Reactions

The discussion on this paper centered on the analogy between the plot of
rate of pressure rise (dP/dt) versus pressure (P) and DTA results.
Dr. Klais had measured most directly the property (P) which he needed to
know.   For example, DTA on the three isomers of nitrobenzaldehyde
suggested different heats of decomposition; but direct measurement of
pressure and temperature showed that in reality there is no difference.

The dP/dt versus P plot did not show autocatalytic decomposition because
this stage was not reached.   It was pointed out from the floor that a
plot of dP/dt versus P cannot produce analagous results to DTA when the
gaseous products are condensing.

FIRE AND EXPLOSION

I. Hymes

UKAEA

Paper 1: Dr. A. F. Roberts

Dr. S.J. Skinner, (Monsanto), asked if there had been any additional tests at Buxton to assess the durability of the various insulation materials used.

Dr. Roberts replied that in this series of trials there had been no such tests and that many properties in addition to thermal insulation needed to be taken into account when assessing the suitability of an insulation system.

Dr. Skinner, in a further question, asked if a steel shell had been used to hold the various insulants to the tank.

Dr. Roberts, who had indicated in his talk that he would be elaborating on the insulation, said that some of the insulation systems had been covered by a steel shell to give weather protection.

Replying to a question from Dr. J.E.S. Venart, (University of New Brunswick), Dr. Roberts said that the choice and size of PRV in the trials had been based upon the standard design requirements for the size of propane cylinder used. Valves of the same capacity had been used throughout the test series and as it emerged that the heat was affecting the performance of the valves, efforts had been made to steadily improve the insulation of the valve.

Dr. Venart then wanted to know if Dr. Roberts had a view on what caused the PRVs at Buxton to lift prematurely, i.e was it damage to the seat or softening of the spring? He would have expected film boiling to commence between 350 and 530 KW/m² yet Dr. Roberts' paper indicated about 250 KW/m². Could the transition to film boiling be due to the PRV performance?

Dr. Roberts replied that at Buxton there were clear indications of when film boiling had occurred and agreed that in some cases the pressure fluctuations when the PRV operated may have triggered the transition. More work to clarify the importance of this effect was probably merited.

Dr. Venart next asked if the percentage fill had been greater than 40% in any of the trials.

Dr. Roberts said the fill had been 40% in each case.

Dr. Venart then explained that he had conducted some experiments - not with propane but R11 and R12 - in which for the particular tank used and heat inputs of > 100 KW/m he had trouble with a two phase discharge from the PRV. Dr. Roberts said that in the Buxton trials as far as one could judge only vapour had been vented and that the tank always discharged in a safe manner.

Dr. I. Swift, (Fauske Associated USA), commented that many types of PRVs had been used in tests on railcars by Salat of the University of Maryland. None of these tests had proved very successful.

Dr. Roberts said he was aware of the work by Salat. In answer to a further question by Dr. Swift about the rate of temperature rise criteria for the insulants used at Buxton, Dr. Roberts said that in his trials the rate was as specified in the ASTM high rise test.

Paper 2: Mr. K. Palmer

Dr. A.V. Howard, (CEGB), noting the use of nitrogen as inert gas in the FRS trials asked whether the alternative use of chlorinated hydrocarbons had been considered. Mr. Palmer replied that alternative suppressants had not been looked for. Cholorinated hydrocarbons would be expensive and whilst their injection to about 10% v/v would suppress flaming, the extinction of smouldering would require the flooding of the test volume.

In a further question Dr. Howard enquired whether coal dust had been used in any of the FRS tests and if so what type of coal had been used and were the vapours analysed?

Mr. Palmer confirmed that coal dust had been used. A soft coal had been chosen for its ready smoking and smouldering characteristics. About 35% m/m volatiles were typically evolved. Analysis of the vapours would be complicated and had not been attempted.

Professor Napier commented that Finch of the BCURA had published some information on the composition of coal volatiles.

Mr. D. Chambers of Babcock Woodall Duckham asked if the FRS had experimented upon PVC cable insulation. Mr. Palmer thought it unlikely that PVC would exhibit the smouldering phenomenon since its unusual smoke gneration mechanism gives rise to large initial volumes of hydrogen chloride. Should the PVC be subjected to gross prior overheating there may be problems but no definitive opinion could be given.

Dr. I. Swift of Fauske and Associates, USA, asked whether any materials had been investigated in the FRS smouldering trials which hadn't given rise to flammable fumes.

Mr. Palmer explained that the aims of the trials had not included this point but certain substances having low calorific values may show such behaviour.

Paper 3: Dr. D.K. Pritchard

Mr. Neville of the British Gas Research Station congratulted Dr.
Pritchard in producing an excellent paper. Commenting that similar
work had been carried out at BGC Newcastle, Mr. Neville asked if in the
Buxton work consideration had been given to attaching a transducer
directly onto the cantilever. Dr. Pritchard replied that it was not
possible to attach transducers directly to the cantilevers used in the
trials. However the good agreement between experiment and theory, for
elastic response, did show that the indirect methods of measuring
loadings he had adopted were accurately measuring the cantilever
loadings.

Professor Napier asked if the current state of the art leads to
any conclusions regarding the accuracy of the early estimates of the
damaging pressure distributions at Flixborough and Pernis.

Dr. Pritchard pointed out that his work was only one part of the
overall picture and had to be used in conjunction with other techniques. The damage caused by any explosion depends on the blast waves
peak pressure, duration and shape. From computer models of unconfined
vapour cloud explosions and experiemntal work a good guess can be made
at the wave shape. Therefore, if an estimate can be made of the peak
pressure at a given loction, for example from some characteristic
damage to a piece of equipment, his model could be used to obtain estimates of the probable duration from damage to cantilevers. Alternatively if
the duration could be estimated the model could be used to obtain the
probable peak pressure.

Dr. Roberts asked if elastic displacement knowledge was sufficiently developed to predict blast effects on e.g distillation
columns and pipework. Dr. Pritchard confirmed that the elastic theory
as described in his paper could be used to determine the response of
pipework and hence check the load on connections between vessels.

EXPLOSION HAZARDS

Rapporteur P. Webster

The Associated Octel Company Ltd.

Paper 1: D.B. Smith, P. Roberts

Dr. Schacke (Bayer) drew attention to work in a spherical vessel
sponsored by a German working party which gave the Coward and Jones
limits, and to the effect the flame propagation criteria adopted have on
reported limits. Dr. Smith considered that the widest observed limits
should be employed for safety applications, bearing in mind the pressure
effects noted in the paper. Wider limits had also been observed by
USBM when varying the source of ignition and in his own work complete
combustion to the wider limits had been observed in stirred gas mixtures.

The need for using a nitrogen purge before air admission to plant
containing ammonia was queried by E. Rigolli (Snamprogetti). Dr. Smith
had not worked with ammonia but cited Russian observations of similar
different types of propagation in large cubes. He suggested that,
despite the limited flammability of ammonia, plant should be purged.
Questioned on the acceptability of purging with fuel, Dr. Smith expressed
some caution; the effects of high pressure in extending particularly the
upper limit, and the possibility with some fuels of cool flames must be
borne in mind. Also any subsequent leakage of air into the system could
negate a fuel purge approach.

Paper 2: M.R. Marshall

Applications of the paper's conclusions to larger tanks were explored in
questions by Prof. Steen (PTB) and Dr. Morrow (South West Research
Institute, USA). In very large tanks, the pattern observed in the test
chamber would probably not be established because of lack of penetration.
No attempt had been made by the authors to analyse the results to allow
scaling, e.g. by Froude No criteria; on the scale of the experiments
buoyancy considerations were overcoming natural turbulence in the
container. Where some heat flux would also be expected, e.g. a leak from
a compressor, some account should be taken of the additional convective
flow.

Dr. F. Evans (National Nuclear Corporation) described some tests on
releases of carbon dioxide at 500°C into a volume containing upper and
lower vents. At this condition it should have positive buoyancy and
tend to pass toward the upper vent. In practice, during the early
stages the gas lost heat to its surroundings and formed a layer on the
floor, much as described in the paper, except that there were significant
concentrations at the upper vent. At a later stage, presumably when the
test chamber was sufficiently warmed, the air/CO_2 mixture contained in it
became buoyant and a new stable condition was attained in which air was
drawn in through the lower vent and all subsequent CO_2 release was vented
from the top, by natural circulation processes. We had anticipated that
natural circulation would occur, but at an earlier stage. This

illustrates that if thermal effects are present, the prediction of concentrations may become considerably more complicated.

In answer to a question by Dr. Sinclair (UMIST), the author suggested that the consequences from multiple leaks could be assessed by summing the fuel inputs.

Paper 3: D.H. Napier

Dr. Nettleton (CERL) pointed out that since, in practice, particles might not be pure it was of interest to know whether small amounts of impurity had a significant effect on the "constant" K.   Dr. Napier agreed that since charging is an interfacial phenomenon, trace impurities could well be of significance, but knew of no relevant work; the fundamental approach of Ref.10 (Davies) perhaps offered a way of investigating the effects of impurities.

Mr. Ling (Roche Products) asked what mechanism might yield the small but measurable charges observed in the centres of fluidised beds.   Dr. Napier thought the most likely cause was inter-particulate collisions rather than gas-solid interactions and involved variations in detailed structure from particle to particle.

Paper 4: K. Schampel and H. Steen

Dr. Lemkowitz (Delft Univ. of Tech.) sought for any qualitative mechanism which might account for the effects of diameter on run-up distance to detonation.   Prof. Steen suggested that a qualitative explanation might lie in a relationship between the scale of cell structure observed in detonations and pipe diameter, but there were no results available to allow a quantitative treatment.   In answer to a further question, Prof. Steen emphasised that all the work reported was in smooth pipes without obstacles; in general, whilst obstacles shorten the run-up time, they do not affect the run-up distance.

In response to Dr. Howard (CEGB), Prof. Steen confirmed that experiments had been also carried out on flowing gases but no significant effect on run-up distance could be detected; he agreed with Dr. Schildknecht that this velocity was small compared with those developing during the transition process ahead of an accelerating flame, nonetheless they fairly represented the flows likely in normal plant operation.

Dr. Bauer (Poitiers Univ.) asked about any effects of energy of initiation.   Dr. Steen acknowledged that whilst very high initiation energies could give rise to a shock wave sufficient to stimulate direct initiation of detonation there was in general no effect of initiation energy on run-up distance in the range studied.

Dr. Elsworth (Shell Research) elaborated the author's further comment that curve 4 of Figure 6 showed evidence of over-driving, leading to a pseudo-detonation which then decayed before ultimately the reaction zone and shock wave recombined to yield a detonation.

Dr. Elsworth commented that whilst ca 50gm of explosive might yield direct initiation of CH_4/air detonations in tube experiments and of spherical detonations in ethane, ethylene or butane, ca 22 kg explosive is required to initiate detonation in an unconfined methane/air cloud.

Dr. Elsworth also referred to the observation of measurable detonation cells in a detonation. Since these cells are large (sic) e.g. for methane/air mixtures, but smaller for other hydrocarbon/air mixtures in which detonations can more readily occur, detonation cell dimensions may offer a basis for understanding the diameter effects on run-up distance observed in the reported work.

Again Prof. Steen acknowledged a qualitative link between the scale of cellular structure and the pipe diameter in which detonations would occur; he knew of no quantitative relationship between cell structure and run-up distance.

Paper 5: M. Nettleton

In opening the discussion, Dr. Burgoyne was interested in any remarks the author might make on the differences between detonation and the shock waves associated with deflagration examined in the paper. Dr. Nettleton replied that the work reported was relevant to the decaying wave associated with deflagration. A detonation wave on the other hand sought to re-establish itself throughout the bend and could give rise to extremely high pressures at some points whereas at others the appropriate C-J pressures are not reached. The highest loads are generally (if not necessarily) generated on the outside of the bend. In reply to Dr. Swift (Fauske and Associates), Dr. Nettleton pointed out that shock diffraction is self-similar and there should be no scale effects on going to much larger ducts; in practical terms, such ducts might present high surface roughness or obstacles but further experiments would be necessary to explore their effects.

Dr. Lamnevik (National Def. Res. Inst., Sweden) had observed with hydrogen/air as fuel, that deflagrations passed into a bed of 3 - 10cm particles in a 25cm diameter steel tube always gave rise to detonation. Dr. Nettleton suggested that the high peak pressures generated by reflections of the blast wave in the system, essentially consisting of many and sharp bends, would give rise to sources of high temperature and pressure and render detonation likely.

Dr. Howard (CEGB) noted that discussion had considered decaying blast waves and constant velocity detonation waves but had not dealt with the possibility of accelerating fronts. Dr. Nettleton observed that in his experiments relevant to either blast or detonation waves, the fronts in the bend were curved and were hence decaying.

Dr. Bauer questioned whether the presence of a bend might influence subsequent expansion. Dr. Nettleton considered that if the structure of the detonation wave was closely spaced, then a bend was likely to have little effect on expansion behaviour; he was not prepared to comment, e.g. on behaviour in fuel-lean mixtures, where this does not hold.

REPLY TO WRITTEN QUESTIONS SUBMITTED TO D.B. SMITH

QUESTION from Mr. T.U. Madhumani, Loss Prevention Association - India

In large loops of hydrogen service, involving pipelines, reactors and heat exchangers, irrespective of quantity of nitrogen used or

time duration of purging, we seem to end up with flammable gas
mixture pockets.   Is there a practical solution to ensure that
such pockets are displaced out of the loop?

ANSWER

This is difficult to answer fully without having further details
of the plant.   But two possible causes of unsatisfactory purging
are: (i) the presence of dead-legs or other irregular shaped
volumes which are by-passed during the purging, and (ii) surface
desorption of the fuel.   Changes in purging procedure may eliminate
the first;   the second seems harder to cope with.

QUESTION from Dr. M.A. Nettleton (CEGB)

Your work has clearly demonstrated the influence of the shape
of the experimental equipment on the determination of flammability
limits.   Are you confident that possible variations in the rate
of energy deposition and amount of energy deposited from the
ignition source will not have a similar effect on limits?

QUESTION from M. Steensma, Akzo Chemie Nederland

You mentioned that in your experiments electrical spark ignition
was used.

For the determination of dust explosion limits we usually apply
"chemical pills" (pyrotechnic mixture, yielding some IOJ).

If you would have taken these chemical "pills" in your experiments
do you think this would have influenced your measurements and/or
conclusions?

ANSWER

We are confident that our limits do not depend in any significant
way on ignition source characteristics.   A number of comments can
be added: (i) the criterion of flammability should be a mixture
capable of self-sustained flame;   (ii) if ignition energies are
too low, limits of ignitibility rather than flammability will be
measured.   We have checked for this by varying our energy;
(iii) ignition sources with high energy or rapid rate of energy
deposition might "overdrive" the flame in its initial stages.
Clearly the greater the energy the larger the influence will be.
But such effects will normally be dissipated within say 50mm of
flame travel.   In our work, flames had to travel 300mm for a
mixture to be deemed flammable.

This point is linked with (i) above.   (v) We cannot envisage
different energy sources promoting other forms of combustion
for the fuels we studied.   Thus we are outside the range of
detonability.   And while it is possible to widen limits on
the rich side by the generation of cool flames (as mentioned in
answer to a previous question), such effects do not occur with
these fuels at atmospheric pressure.

WRITTEN QUESTION FOR H. STEEN FROM D. CHAMBERS, BABCOCK-WOODALL-DUCKHAM

Fig. 2 for 50mm Pipe gives minimum S_D of $\simeq 4.5$m for Propane

Fig. 3 gives $\simeq 7.5$m

Could you comment on this apparent discrepancy?

ANSWER FROM H. STEEN

The experimental results indicated in Fig. 2 are obtained under slightly different test conditions (e.g. blockage of pipe end) compared with the results in Fig. 3 and are therefore not directly comparable in a quantitative manner. The experiments related to Fig. 2 were carried out just to demonstrate that the most incendive concentration gives the smallest run-up distance.

WRITTEN QUESTION FOR DR. M.A. NETTLETON, CEGB TO PROF. STEEN

In my reading of the literature the run-up distances in tubes of small diameter (S_D) obeys a law of the form $S_D \propto \sqrt{D}$. Your results appear to show $S_D \propto D$. Can you comment on what you would consider the transition diameter for this change in dependence?

ANSWER FROM H. STEEN

Our experiments refer to diameters of 50mm and more. With significantly decreasing diameter the flame movement and acceleration will be more and more influenced by wall effects (boundary layer). This is to my mind a plausible interpretation of the fact that the dependency is not linear for low diameters but weaker. The limit diameter depends on the flammable gas of course and should be between about 25mm (for CS_2, see ref.12) and about 50mm (for propane) according to our experiments.

EXPLOSION RELIEF

R.J. Harris

British Gas Corporation

Paper 1: Performance of Low Inertia Explosion Reliefs Fitted to a 22m
Cubical Chamber.
P.F. Thorne, Z.W Rogowski and P. Field (Fire Research Station,
UK).

Discussion on this paper was concentrated on factors surrounding
the generation of oscillatory pressure peaks. In reply to J.A. Eyre
(Shell Research-UK) who asked why the highest pressures in the experiments occurred with rear ignition (Figure 6 of the paper) Mr. Thorne
indicated he could offer no precise explanation. He explained that
what was needed was a physical picture of what was happening and
suggested that useful information might be gained by taking high speed
photographic records within the test chamber.

E. Skramstad (DNV-N) said that DNV had carried out similar experiments with propane to those described, and that the results agreed with
those in the paper. He added that the DNV results showed that
acoustic pressure peaks were more pronounced in rich mixtures, that
their occurence was sharply concentration dependent and that they had
obtained high peaks with both rear and central ignition. He asked if
Mr. Thorne had any explanation for the concentration dependence. Mr.
Thorne replied that he knew of other work which had shown this concentration dependence but that the work described in his paper offered
no explanation.

M. Schildknecht (Batelle Institut-D) commented that he was asto-
nished at the influence of rock wool in damping the oscillatory
pressure peaks as described in the paper. In comparing the overall
form of the two pressure traces (with and without rock wool) he was
surprised that the time scales of the two events were nearly the same.
Mr. Thorne enquired whether M. Schildknecht was asking whether the
acoustic peak was a real pressure. He said he knew of experiements
which had proved it to be so. In these experiments two vents, one
large, one small, were installed in a single chamber. The larger vent
was blown out early on and it was not until very late on in the explosion (coinciding with the acoustic peak) that the smaller vent panel
was ruptured.

Paper 2: Dust Explosion Experiments in a Vented 500m Silo Cell
R.K. Eckhoff and K. Fuhre (CMI-N)

T. Milsom (European Risk Management) asked whether in the light of
the results from the experiments described, the authors had yet proposed any modifications to the guidelines on dust explosion venting
issued by the code setting bodies. Dr. Eckhoff replied that they had
not yet done so. He said however that we should recognise that the
present guidelines reflected the results obtained from particular
experiments.

New experiments were showing that we could not rely on the idea that
the combustion rate in dust clouds was purely dependent upon the dust
type. He believed that greater attention should be given to the
influence of the distribution of dust concentration and any initial
and explosion induced turbulence. He thought that large scale
experiments should be encouraged.

E. Skramstad (DNV-N) pointed out that the results given in the
paper showed a strong correlation between the maximum pressure developed and the time at which the flame appeared at the vent opening.
He said that work at DNV suggested that this correlation could be
attributed to the onset of the venting of burnt gases and Dr. Eckhoff
agreed that this was the most likely explanation.

Paper 3: Gaseous Combustion Venting - A Simplified Approach
I. Swift (Fauske and Associates-USA)

Dr. Swift was asked by M. Nettleton (CERL-UK) how sensitive
results from the model were to the choice of time dependent variables
and which of the variables was most important. Dr. Swift said that
both the burning velocity and vent area (which has a dependence on the
vent opening time) had a time dependency. Burning velocity values were
corrected at each time step for their dependence on pressure and temperature. However, he pointed out that the results produced were most
sensitive to the choice of the value of the discharge coefficient.

M. Schildknecht (Batelle Institut-D) asked if the turbulence properties of the gases flowing towards that vent opening were calculated
from the mean flow velocity. Dr. Swift replied that the fluctuating
velocity was taken to be 5% of the mean flow velocity in the vessel
during venting.

Dr. Lewis (Consultant-UK) noted that in tables 1,2 and 3 in the
paper, calculated pressures were consistently in excess of the experimental results against which they were compared. He asked whether
calculations had been done to see by how much the vent areas as used
in the model would have been increased in order to exactly predict the
experimental results, and if so whether these calculated areas could
be used in practice. Dr. Swift replied that these calculations had
been done. Predictions from the model were however conservative. He
pointed out that nomograms were given in the paper which would allow
vent areas to be calculated in order to restrict the maximum pressures
generated to a given level for a range of vessel volumes. W. Howard
(Consultant-USA) asked how these nomograms compared with others that
had been published. Dr. Swift replied that the nomograms gave similar
results to those produced by Bartknecht, except for larger vessel
volumes in which case nomograms given in the paper predicted a
requirement for larger vent areas to restrict the pressure to a given
level.

Finally, M. Schildknecht asked whether the model presented,
allowed for volume scaling since he believed it was known from experiments that the peak pressures developed depended upon the size of the
vessel. Dr. Swift indicated that his model predicted the same maximum
adiabatic pressure for any vessel. He noted however that heat losses
were not included and pointed out that these would differ between
large and small vessels.

Paper 4: Venting of Gas Explosions in Large Rooms
C.J.M. Van Wingerdern and J.P. Zeeuwen (TNO-NL)

R. Eckoff (CMI-N) asked to what extent the results given were applicable to practical situations – the experiemnts described were carried out in empty enclosures and it was known that obstacles could enhance burning rates through the effects of turbulence. Dr. Zeeuwen replied that the experiements described were designed to study acoustic oscillations. Since other work had shown that obstacles inside a vessel could interfere with acoustic disturbances the tests were not carried out with obstacles. He also considered that in practice many rooms were nearly empty and in terms of turbulence generation one or two obstacles would not be important. He noted that scaling of the effects was a difficulty because some instability effects were scale dependent. There was no way of knowing whether other instabilities would come into play with larger volumes.

In reply to A. Duxbury (ICI-UK) who asked whether acoustic pressure peaks could occur with dust explosions, Dr. Zeeuwen noted that in the case of dusts the flame was thicker, conditions were not really the same, and he didn't believe they would be important.

I. Swift (Fauske and Associates-USA) asked whether the provision of vent openings on opposite walls of a vessel would reduce or eliminate acoustic peaks, and if so whether this could be a better practical solution that lining the vessel walls with acoustic absorbant material. Dr. Zeeuwen said that although they had not carried out tests to check, he believed that provision of vents on opposite walls would have this effect. He therefore believed that this was a possible practical alternative means of suppressing acoustic peaks. He pointed out however that in practice it was often not possible to place vents in the positions required.

J. Eyre (Shell Research-UK) asked if there was any explanation for the observation that in some tests acoustic oscillation seemed to add together, suggesting increases in burning rate, whilst in others this was not so pronounced. Dr. Zeeuwen replied that increases in burning rate were due to coupling between the flame and acoustic disturbances and that this was most efficient in mixtures which were inherently unstable. He believed that small variations in gas concentration could be crucial. In addition he pointed out that for coupling to occur acoustic waves had to be produced first and that there may be variations in the time scale between experiments meaning that acoustic disturbance – flame coupling is less efficient in some experiments than others.

E. Skramstad (DNV-N) asked whether the burning taking place outside the vent opening was influencing the burning inside during experiments. He had noted in the slides shown in the presentation that the flame outside the vessel appeared more intense in experiments in which acoustic oscillations occurred. Dr. Zeeuwen said he was not convinced that the burning outside was having any great influence on that inside, since the origin of the acoustic waves could be traced back to a time at which the explosion outside the vessel was not luminous.

Finally, W. Howard (Consultant-USA) made an appeal for different international research programmes to be co-ordinated if at all possible. He believed that this was the only way by which better progress could be made towards developing a rigorous and realistic mathematical model of this complex problem.

SPECIAL DESIGN PROBLEMS

Rapporteur Mr. M. Kneale

John Brown Engineers & Constructors, B.V.

In the first paper, 'Protection of distillation columns against vacuum by loss of heating', R. Gill (Scientific Design) stated that the system described using nitrogen injection is quite often used. They have found that if the nitrogen is injected close to the condenser much smaller amounts are effective.

Care has to be taken to avoid a negative pressure gradient up the column particularly if valve trays are used.

The 15 second response time allowed in the table seems very adequate. Valves with response times between 200 milliseconds and 1 second are available so this should not be a problem.

The author replied that the paper illustrated a solution to a particular existing plant design where nitrogen flow through the column was selected.

Then J. Chapman (John Brown E & C) asked whether it would have been better to stop the flow of overheads. There are fast response valves available. The introduction of nitrogen as loss of reboiler will initiate shutdown in any case. Are there not shutdown valves in the overhead system which can fail erroneously in any case?

Mr. Fitt answered that this existing plant had no overhead valves and these very expensive items were not recommended.

In the second paper, 'Deterioration of structures supporting refinery equipment', A. Perry (Caltex) said that it was emphasised that chlorides cause problems. What about refinery structures actually submerged in sea waters?

Mr. B. N. Pritchard answered that for marine applications, measures must be taken at the design stage. Basically there are two useful measures:

a) the use of very dense concrete;
b) the use of sulphate resistant cements;

Dr. S. J. Skinner (Monsanto) asked: "Is there relevant experience with lightweight and fireproofing concretes?" In reply, the author said they advocate painting the steel and then inspecting paint and steel, under the concrete regularly.

In reply to the question: "Have you any experience with deliberate addition of chlorides to concretes?", the author responded by saying that calcium chloride was used during the 1950's to improve the setting period. The results were disastrous as is well known.

Finally, R. Salter (J. H. Minet & Co) asked if there is a problem with corrosive attack proceeding beneath grade (below ground). The author answered that the problem is not so severe probably because air is relatively absent. The chloride concentration is usually also lower.

After the final paper to this session had been presented, 'Determining safety zones for exposure to flare radiation', B. N. Pritchard (Esso Engineering Europe) asked "How did you achieve the high Reynolds numbers of 10^3 and 10^5 quoted in the paper in a wind tunnel?" The reply given was "Where the flare diameter of 1 mm Re of 10^4 was achieved."

The key question in any model is value attached to the fraction of heat released? Have you used the same fraction as recommended in API? If you use a solid emitter as a flame model we find at 1 to 1½ flame we have very good agreement with API. We believe your model underpredicts radiation.

In reply, the author said that he believed the Holme-Luce prediction leads to overdesign. The main reason is that Holme-Luce allows too much for flame buoyancy.

J. W. Hempseed (Air Products) said that the wind tunnel stack had a diameter of 1 to 4 mm and a height of 100 mm. What problems do you see in scale up to plant size by factor of 100 or more? Did you measure existing plant?

Prof. G. Fumarola said that this is the real problem. Usually scale up is done by (constant) Reynolds number. This is, of course, an approximation. We did not measure existing flares.

R. Mill (Exxon Chemical Co, USA), the Chairman of the session, commented that while the Brzustowski model may be somewhat conservative we have found it acceptable and useful. We tend to define practical rules such that an operator has not more than 6 seconds to get out from the area under the flare. In practice this is equivalent to a radiation level of 9.45 KW/m^2 maximum.

The final question by D. Chambers (Babcock Woodall Duckham) was addressed to B. N. Pritchard: "What is the effect of flare radiation on structures? What are the temperatures that common structures can withstand (before we should get worried?)

He replied that there is no problem with steel up to $300^{\circ}C$. Concrete is only happy up to $80\text{-}90^{\circ}C$. The most severe problem arises if the structure is very stiff. A stiff structure will be overstressed by very small temperature changes.